The Lysenko Effect

CONTROL OF NATURE

Series Editors

Morton L. Schagrin
State University of New York at Fredonia

Michael Ruse
Florida State University

Robert Hollinger
Iowa State University

Controlling Human Heredity:
1865 to the Present
Diane B. Paul

Einstein and Our World
David C. Cassidy

Evolutionary Theory and Victorian Culture
Martin Fichman

The Lysenko Effect: The Politics of Science
Nils Roll-Hansen

Newton and the Culture of Newtonianism
Betty Jo Teeter Dobbs and Margaret C. Jacob

The Scientific Revolution:
Aspirations and Achievements, 1500–1700
James R. Jacob

Scientists and the Development of Nuclear Weapons:
From Fission to the Limited Test Ban Treaty, 1939–1963
Lawrence Badash

Technology and Science in the Industrializing Nations, 1500–1914
Eric Dorn Brose

Totalitarian Science and Technology
Paul R. Josephson

Nils Roll-Hansen

The Lysenko Effect
The Politics of Science

WITHDRAWN

Humanity
Books

an imprint of Prometheus Books
59 John Glenn Drive, Amherst, New York 14228-2197

Published 2005 by Humanity Books, an imprint of Prometheus Books

The Lysenko Effect: The Politics of Science. Copyright © 2005 by Nils Roll-Hansen. All rights reserved. No part of this publication may be reproduced, stored in a retrieval system, or transmitted in any form or by any means, digital, electronic, mechanical, photocopying, recording, or otherwise, or conveyed via the Internet or a Web site without prior written permission of the publisher, except in the case of brief quotations embodied in critical articles and reviews.

Inquiries should be addressed to
Humanity Books
59 John Glenn Drive
Amherst, New York 14228-2197
VOICE: 716-691-0133, ext. 207
FAX: 716-564-2711

09 08 07 06 05   5 4 3 2 1

Library of Congress Cataloging-in-Publication Data

Roll-Hansen, Nils, 1938–
 The Lysenko effect : the politics of science / Nils Roll-Hansen.
   p. cm. — (Control of nature)
 Includes bibliographical references (p.   ) and index.
 ISBN 1-59102-262-2 (pbk. : alk. paper)
 1. Lysenko, Trofim Denisovich, 1898-1976. 2. Biology—Research—Soviet Union—History. 3. Genetics—Research—Soviet Union—History. 4. Geneticists—Soviet Union—Biography. 5. Plant breeders—Soviet Union—Biography. 6. Agriculture and state—Soviet Union—History. 7. Science and state—Soviet Union—History. 8. Communism and science—Soviet Union—History. I. Title. II. Series.

QH31.L95R65 2004
570'.72'047—dc22

2004012970

Printed in the United States of America on acid-free paper

Contents

Preface	9
1. Science Policy and the Problem of Lysenkoism	11
2. The Situation in Agricultural Plant Science	19
Classical Genetics	21
Genetics and Plant Breeding	23
Developmental Physiology:	
Effects of Light and Temperature	28
The Imperial Legacy of Soviet Agricultural Research	32
Soviet Plant Physiology around 1930	36
Soviet Plant Breeding in the 1920s	38
Nikolai Vavilov—Entrepreneur of Science	40
Anti-Mendelism in Soviet Biology	46
3. The "Barefoot Professor"	53
Lysenko's Start as a Plant Breeder	53
The "Barefoot Professor"	56
Vavilov Takes an Interest	58
Temperature Effects on Plant Development	58
Leningrad Congress, January 1929	64
From Wishfulness to Cynicism	69

CONTENTS

4. Marxism and Science Policy	75
Planning of Science	77
Philosophical Ideas behind Science Planning	78
Biology and Philosophy	84
"Creative Darwinism"	86
The Lenin Academy of Agricultural Science (VASKhNIL)	89
Revolution at the Institute for Plant Industry (VIR)	92
Reorganization of VASKhNIL, 1934–1935	95
Implementing the "Unity of Theory and Practice"	100
"Socialist Competition" in Science	102
5. Vernalization	113
Lysenko's Success in the Mass Media	114
Recognition and Criticism from the Leading Scientific Experts	116
Promoting the Vernalization of Seed Grain	120
Criticism and Abandonment of Vernalization	126
Agriculture in the Far North	131
Enthusiasm among Botanists	133
Theoretical Problems of Vernalization	140
The International Response	142
Summing Up	148
6. From Problems of Plant Breeding to Controversy in Genetics	155
Vavilov's Science of Plant Breeding	155
Vernalization as a Tool for Plant Breeding	159
Lysenko Turns to Breeding and Genetics	163
The June 1935 Meeting in Odessa	169
New Wheats in Record Time	174
Dialectical Materialism in Genetics	178
Vavilov's Role	184
7. "Two Directions in Genetics"— The Congress of December 1936	193
Lysenko versus Vavilov	197

Counterattack of the Geneticists—	
Enter Kol'tsov and Muller	202
Shades of Sympathy for Lysenko	206
Summing Up: Meister's Compromise and	
Muralov's "Basic Results"	209
Politics behind the Scenes	213
8. Against Bourgeois Ideology in Science	**217**
Kol'tsov in the Aftermath of the	
December 1936 Congress	218
The VASKhNIL *Aktiv*, March 26–29, 1937	221
Cancellation of the International Congress of Genetics	230
9. Lysenko Takes Over	**249**
Lysenko Becomes President of VASKhNIL	250
Role of Philosophy—1939 Conference	252
Lysenko Takes Control of Genetics	262
Epilogue: From Temporary Optimism	
to Final Rout of Genetics	265
10. Concluding Discussion: Why Did It Happen?	**281**
Autonomy of Science	282
Western Reactions and the Feedback	
to Soviet Politics of Science	286
Explaining the Rise of Lysenkoism	290
Lamarckism	294
Impact of Marxist Philosophy of Science	296
Abbreviations and References	303
Index	325

Preface

This book has taken a long time to finish. It started some time in the early 1970s when I taught a course in philosophy of biology with Jon Elster. One day he said, "You know Russian and you know plant physiology, so why don't you explain this Lysenko business to us?" The first result was a never-published paper called "The Lysenko Affair: A Failure in Science Policy" (in Norwegian). A large part of the work was carried out between 1977 and 1993 when I was employed at the Norwegian Institute for Studies in Research and Higher Education (NIFU). In 1982–83 a Hambro fellowship to Clare Hall, Cambridge, England, gave me the chance to get acquainted with the archives of the Imperial Bureaus of Plant Breeding and its collection of Russian agricultural science journals from the interwar period. A book in Norwegian, *Wishful Thinking as Science, The Rise of Lysenkoism in Soviet Agricultural Science, 1928–1938*, based on printed material and secondary sources only, was published in 1985. The Wissenschaftskolleg zu Berlin provided excellent working facilities in 1987–88. During the 1980s and early 1990s I made many trips to make use of Russian archives. My colleagues in Russia were indispensable in guiding me to relevant documents. During a longer visit in 1986, Luisa Frolova and Evgeniia Senchenkova provided a number of contacts with researchers in plant physiology and agricultural science who had been involved in the controversies over

Lysenkoism. Vassili Babkov and Kirill Rossianov were particularly helpful with the Moscow archives, and they have also been important as discussing partners doing research on the history of Soviet genetics. Kirill kindly read through the manuscript at a late stage, giving much useful advice. Tania Lassan was very helpful with information on VIR, its literature, and its archives. Olga Elina read and commented on the agricultural science chapters. Daniel Aleksandrov provided essential help in the St. Petersburg archives. Together with Alexei Kojevnikov, he has also been a good guide to general themes and questions of the history of Soviet science.

An enduring project like this is naturally a considerable load on the closest family. I am grateful to Ingeborg Glambek for her stamina and good advice in difficult situations.

—Nils Roll-Hansen
Department of Philosophy, University of Oslo
October 5, 2004

Chapter 1.

Science Policy and the Problem of Lysenkoism

The more closely science is involved with politics and practical affairs, the more pervasive becomes its dependence on social goals, values, and wishes. On the other hand, excessive autonomy threatens with isolation and a narrow-minded search for socially irrelevant truths. This dilemma was very much on the minds of politicians and the public as well as scientists in the optimistic and revolutionary era of the Soviet Union. At the beginning of the twenty-first century it is again a central problem both in practical science policy and in theorizing about science.[1]

Traditionally, Lysenkoism has been explained as a result of Stalinist despotism, the willful intervention of arrogant political bosses into science. I will attempt a reinterpretation showing how the course of events depended on the participants' views on scientific and philosophical issues. No doubt harsh oppression and widespread terror was the background for everything that happened in the Soviet Union in this period. But to properly understand the irrationality of Lysenkoism and to explain why one scientific theory was preferred over another, a close analysis of scientific and science policy issues is needed. And the policy issues have a philosophical basis that is no less interesting and significant at the beginning of the twenty-first century than it was in the middle of the twentieth.

A few facts to set the historical stage: T. D. Lysenko, an agronomist of peasant origin, born in 1898, was the leader of an agrobiological school that rejected standard (classical) genetics. In 1938 he was made president of the Lenin Academy of Agricultural Science, and in 1948 genetics was officially condemned as unscientific by the Soviet government. Not until 1964 did this suppression end. The background for Lysenkoism was a dramatic period in Russian history: the 1917 revolution; the first five-year plan of 1928–1932 with collectivization of agriculture, forced industrialization, and cultural revolution; the great terror of 1936–1938; and the great war of 1941–1945.

The Lysenko episode had a major impact on the ideological front of the cold war starting in the late 1940s. It has been characterized as the biggest scandal of twentieth-century science. That it happened under a regime that took particular pride in building its policy on science gives reason to reflect. The Soviet Union was the first country with a government policy and large-scale public support for science (Korol 1965; Bailes 1978; Graham 1998). And agricultural science was a main showcase for this unprecedented investment in science. It is an experiment worth studying. Why did the policy of Soviet biology go so wrong?

According to the eminent American historian of Soviet science David Joravsky, it is a romantic Western myth that Marxist theory and scientifically outdated Lamarckian ideas about heredity were important causes of Lysenkoism. In his view Lysenkoism "rebelled against science altogether. Farming was the problem, not theoretical ideology, all the sciences that impinge on agriculture were tyrannically abused by quacks and time-servers for thirty-five years" (Joravsky 1970, ix). Another leading American historian of Soviet science, Loren Graham, agrees that Lysenkoism had little to do either with Marxist theories about science or with serious scientific issues in biology. The "Lysenko episode was a chapter in the history of pseudoscience rather than the history of science" (Graham 1972, 195).

It is true that Lysenko ended up rejecting sound science. But he was not simply a pseudoscientist from the start. Some of his early results were acclaimed by leading Russian scientists as having great theoretical as well as practical interest. They It also attracted lively international interest. A sign of his influence is the term "vernalization," which is still standard in plant physiology for specific phenomena of temperature influence on plant development. The interesting problem is what took Lysenko and so much of Soviet biology into the sphere of what Graham calls "pseudoscience."

An alternative to this totalitarian political explanation was sketched by supporters of the radical science movement. Two well-known American biologists, Richard Lewontin and Richard Levins (1976), saw Lysenkoism as "an attempt at scientific revolution"—a genuine attempt to transform science into a better instrument for social progress and justice. To understand why it failed, they wanted to distinguish its valid principles from the circumstances and misunderstandings that derailed it. In their view Lysenkoism raised important issues about the relationship of theoretical science to practical work that remained unresolved—and still are, we could add, today. As practicing biologists they also found Marxist philosophy to be fruitful in their own research. In other words, the methodological and science policy issues could not be as easily dismissed as the totalitarian explanation assumed.

This Marxist dialectical perspective was not developed beyond the level of a sketch.[2] But it can at least serve to remind us that well beyond the Second World War the sociopolitical project of the Soviet Union was not only based on terror and suppression, it was also carried by a genuine and broad enthusiasm. Especially among scientific and technological cadres this optimism was strong. Even Andrei Sakharov, the atomic physicist and later leading dissident, felt a pang of sorrow at the death of Stalin in 1953. In a private letter he wrote: "I am under the influence of a great man's death. I am thinking of his humanity." Stalin still symbolized the hope of creating a better society and a more peaceful world (Sakharov 1990, 164).

A new generation of Russian historians of science writing in the post-Soviet period have emphasized normal social features of science under Stalin's regime. The complex integration of scientific and political establishments implied an intimate two-way relationship rather than simple subordination of science to politics.[3] Scientists were politically controlled but could also use the political leverage of the system in competing with each other. Soviet science was not unique in this respect. But the monolithic centralized and brutal nature of the regime made the stakes higher and the set different. From this institutional perspective Lysenkoism was a fashion that affected all of Soviet science, and the Lysenko episode becomes a key to its history. However, this institutional approach also has little interest in the role of scientific issues and theories about the nature of science. The early success as well as the later failure of Lysenko and his followers are mostly described and explained in terms of ritualized public "discussions," lobbying of key political bosses, and so on.[4] The social mechanisms driving "Stalinist science" are much the same in the traditional totalitarian and the new institutional accounts. But strong moral condemnation has given way to a more relaxed and relativistic attitude toward the story.

While the cold war lasted, in the 1950s, 1960s, and 1970s, and even in the 1980s, it was natural to focus on the tyrannical character of the Soviet regime. This is, no doubt, an essential aspect of any history of Lysenkoism. But even under tyrannical regimes, reason is a feature of human nature. The tyrants themselves often give reasons for their acts. A problem with most of the earlier historiography of Lysenkoism is that the shocking experiences of the totalitarian "Stalinist" system have tended to obscure the scientific and rational aspects of the Lysenko episode. In the hindsight of the late Soviet period, the struggle to get rid of Lysenkoism loomed large and the conflict appeared primarily as one between science and politics, between rational scientific argument and political power play. From the late 1930s the fronts appeared as clearly drawn, with proper science on one side and politically supported "pseudoscience" on the other. The genuine dilemmas and hard choices of scientific rationality, method, and policy applications that faced the earlier period disappeared from sight. During the rise of Lysenko, culminating with his presidency of the Lenin Academy in 1938, it was not equally obvious where the future of science lay.

Two interesting and informative accounts are *The Rise and Fall of T. D. Lysenko* (1970) by the Russian biochemist and historian of science Zhores Medvedev and *Lysenko and the Tragedy of Soviet Science* (1994) by the Russian geneticist Valerii Soyfer. Medvedev's book played an important part in the internal fight to liberate Soviet science from Lysenkoism. It went through a long development as an underground text in the Soviet Union before it was published in the West. Medvedev is more sensitive than Joravsky and Graham to the scientific credentials of Lysenko and to the support he had within the Soviet scientific establishment. But his book is nevertheless focused on the long struggle to free Soviet genetics from the politically supported oppression under Lysenko. For him as well it is primarily the totalitarian character of Stalinism, the "cult of personality," that explains the phenomenon of Lysenkoism. Soyfer's account also belongs to the totalitarian tradition. Like Medvedev, the fight against Lysenkoism formed his scientific and intellectual career. As a young scientist Soyfer discovered the falsity of what had at first appeared as proper scientific claims, and he joined the difficult and dangerous fight. In his historical account strong emphasis on personal psychology tends to obscure the role of genuine scientific issues and to cover up central questions of historical explanation. For instance, he claims that Lysenko's "very first work was based on deceit and was not at all the 'discovery' that the newspapers trumpeted and that many naively assume to be the case even today" (Soyfer 1994, 299). Certainly newspapers then, as now, exaggerated and often advertised trivialities as revolutionary discoveries. But a closer look at the scientific debate and literature shows that Lysenko's scientific work cannot be brushed off as easily as Soyfer does.

The effect of science policy on the rise of Lysenkoism has special interest because of the Soviet Union's international pioneering role. The Soviet Union was the first country to form a comprehensive public policy for scientific research and technological development. Theoretical inspiration came from socialism and Marxist ideas on the social nature of science. From the early 1920s the Soviet regime was leading the way in government funding and stimulation of scientific research. Efforts were vigorous from the beginning. While scientific and technological utopianism was an important cultural element in the West, it was a state creed in the Soviet Union. At first the efforts were uncoordinated and pluralistic, but with Stalin's great break of 1928–1932 a unique centralized system for planning and managing scientific research took shape. Aiming to overtake and surpass the capitalist West economically as well as in social welfare, Soviet Russia invested more resources in research than any other country (Korol 1965; Bailes 1978).

This deliberate promotion of science for the benefit of society created sympathy and support among liberal and left-wing Western scientists and intellectuals for the Soviet cause in general as well as for similar science policy at home. Theorizing about science policy in the West from the 1930s on was heavily influenced by the Soviet example and by Marxist theory.[5] The Soviet Union developed a system for scientific research and development larger than any other country in the world. Science and technology was the means by which it hoped to win the race with Western capitalism. In the 1980s, just before its unexpected collapse, the number of scientists and engineers relative to the total population was about twice as high in the Soviet Union and its Eastern European satellites as it was in the United States and Western Europe (Graham 1998).

Looking back from the early twenty-first century, the Soviet research system appears highly overgrown and inefficient. We now easily forget that it also produced some impressive successes, like the quick development of nuclear weapons after World War II and the launching of the first Earth satellite, the *Sputnik*, in 1957. The arms race of the cold war was a competition in scientific and technological development. Fear of Soviet leadership in space technology inspired more forceful governmental science policy in the West. Thus Soviet achievements and ideas shaped Western science policy and the Western system of scientific research and development toward the end of the twentieth century. Since the fall of the Soviet Union, Western science policy has been driven as much by a race for economic dominance as for military supremacy—much like it was during the early decades of the Soviet regime.

Besides medicine, agriculture has been the main field of applied biology. Agriculture received much more political and scientific attention in the first half of the twentieth century than it does today. The Malthusian problem of feeding an exponentially growing world population was felt to be pressing, and a much larger proportion of the population was working in agriculture. Russia after the 1917 revolution was still industrially underdeveloped, and the Soviet project of modernization demanded a rapid transformation of the agricultural economy. Stalin's great break starting in 1928 gave science a key role in making agricultural production more efficient. In return for generous funding, agricultural research was to make the forced collectivization of the peasants an economic success.

The Lenin Academy of Agricultural Science (VASKhNIL),[6] established in 1929, marked the special Soviet emphasis on agricultural science. During the Stalinist drive for industrialization from the late 1920s on, issues of agricultural politics and production periodically dominated Soviet public debates. For the party and the government, the Lenin Academy of Agricultural Science appeared as a rival and an alternative to the traditional Academy of Sciences, through its more direct practical relevance. Under the scientific leadership of Nikolai Vavilov, an agricultural plant scientist of high international reputation, VASKhNIL formed the research system where Lysenko made his career. At first Vavilov acted as Lysenko's patron and supporter, but he later became his main opponent and finally his victim, dying in a prison in 1943.

In contrast to the totalitarian interpretation, I will focus on the scientific issues as they evolved in the debates among scientists, administrators, and politicians where the policy for agricultural science was worked out. I will show how the early work of Lysenko, in spite of weaknesses that are easy to see in retrospect, falls within acceptable standards of the time. Some of his ideas and experimental results in plant physiology were even praised by leading Soviet biologists as highly important and interesting. They also attracted broad international attention and remained respected long after his genetics had been ridiculed. The rise of Lysenkoism can no longer be seen as simply the result of illegitimate political interference. Lysenkoism appears rather as an extreme form of tendencies that are inherent to any modern science that is closely bound up with practical political and economic purposes.

My account will emphasize science policy as the factor that links developments in scientific knowledge to the broader economic and political context. To understand the motivations and the reasons of the actors, I will make use of materials from the extensive and intense debates on how science should be constructed and governed under the Soviet regime. There exists a vast amount of published literature in newspapers and more or less popular scientific journals. I will also make use of unpublished verbatim reports from the numerous meetings discussing issues at the science-politics interface in the crucial period of the 1930s. The content of these materials have so far not been much exploited in the historiography of Lysenkoism.

NOTES

1. For recent discussion of the politics of science and the tension between social good and scientific autonomy see, for instance, Michael Gibbons et al. (1994) and Philip Kitcher (2001).
2. The American historian of science Robert Young made a start (1978) that was not followed up. The French philosopher Dominique Lecourt (1977) also interpreted Lysenkoism from an internal Marxist theoretical and political perspective, but only briefly sketched the history on the basis of secondary sources.
3. See, for instance, Kojevnikov (1996, 1999, 2004) and Krementsov (1997).
4. See, for example, Krementsov (1997, 58–60).
5. An influential example is J. D. Bernal's classic *The Social Function of Science* (1939). This book experienced a significant revival with the student revolution of 1968. For the Soviet and Marxist impact on British scientists and intellectuals see P. G. Werskey (1978).
6. In Russian: Vsesoiuznyi Akademii Sel'skokhoziastvennykh Nauk Imeni Lenina (The Lenin All-Union Academy of Agricultural Science).

Chapter 2.

The Situation in Agricultural Plant Science

During the second half of the nineteenth century, countries in Europe and North America made large investments in agricultural research. Agriculture was considered a key factor in modernization. This development was delayed in Russia. But from the 1890s there was rapid expansion of research. Plant breeding was a particularly active area internationally. The first state-supported Russian institutions for plant breeding were well established before World War I, and ambitious plans for an imperial network were set in motion. The main models were leading institutions in Germany and Scandinavia. In particular, the Svalöf Plant Breeding Station in southern Sweden served as a model in scientific as well as practical agricultural and commercial respects (Elina 1997a and b; Lassan 1997).

Roughly, there are two ways of improving the quantity and quality of plant products: better methods of cultivation and better plant stocks. One can either create more favorable environmental conditions for the plants or improve their inherited properties. These two ways correspond to two main branches of plant science, physiology and genetics.

The struggle over Lysenko's teaching (agrobiology) from the mid-1930s on was centered on genetics. But one particular branch of plant physiology was crucial, especially in the early development of Lysenko's ideas, namely, the physiology of development. This branch studies the factors that govern the life cycle of the plant from seed through youth and maturity to a new generation of seed. Both genetics and developmental physiology belonged to the new experimental trend that came to dominate natural science from the late nineteenth century on. Both are typical sciences of the twentieth century. The start of modern genetics is conventionally taken to be the rediscovery of Gregor Mendel's laws in 1900. Developmental physiology is an even younger discipline. Not until the 1920s did the effects of low temperatures, length of day, and so on attract broad interest and sustained systematic inquiry.

The close relationship of developmental physiology to genetics is an important background for our story. Developmental physiology studies the life phases of the individual plant as they unroll under the double influence of environmental factors and genetic instructions. It corresponds to embryology in animal studies. Plant breeding is an applied science in which the two theoretical disciplines of developmental physiology and genetics come together. Trofim Lysenko started his professional career in the early 1920s as a practical plant breeder.

Both genetics and the developmental physiology of plants were very young disciplines when Lysenko started his professional career. But even if genetics was still immature and controversial, it was nevertheless well established as a prestigious specialty by 1930. Geneticists were the young turks of biological science, especially in the Soviet Union. By comparison, the developmental physiology of plants was a peripheral field of research. It attracted a considerable amount of general attention and became a side interest of many plant physiologists in the 1920s and 1930s. But only a few cultivated it as their main field of research. Theoretically, the field was very open. When Lysenko made his most significant contributions around 1930, there still existed no well-established general theory. But the potential for important application of the new science to agriculture was recognized. Research expanded rapidly, especially in the Soviet Union, but also in western Europe, North America, and other countries with an interest in agricultural research.

CLASSICAL GENETICS

The new discipline of genetics attracted great public interest from the beginning. Theoretically it promised a new unifying view of the living world, and the practical possibilities appeared no less impressive. Many people thought it likely that soon evolution, not only of plants and animals, but also of man, would come under human control. Besides plant breeding, eugenics, that is, practical policies to improve the hereditary quality of human populations, was the main source of inspiration and funding for the development of general genetic theory. From the late 1920s eugenics became increasingly suspect because of its connections to reactionary and fascist politics. The historical links to eugenics then proved fateful for genetics and indirectly also for plant breeding in the Soviet Union.

Roughly, one can say that genetics grew out of evolutionary studies as their experimental branch. Genetics was created by the quest for an explanation of evolutionary change. The hard core of classical genetic theory was the assumption of hereditary factors, soon called "genes," that are transferred from one generation to the next. The genes are in general stable, but change occasionally in a stepwise way by so-called mutation. This double feature of stability and change provides a suitable basis for the explanation of biological evolution. The term "gene" was first coined by the Danish plant breeder and geneticist Wilhelm Johannsen (1909, 124). It was the demonstration that such stable factors, genes, actually exist that got genetic science off the ground.

In classical genetics there are two ways for the genetic makeup of an organism to change from one generation to the next, namely, mutation and recombination. Mutation means a change in one individual factor. This change is reproduced as the factor is copied and transferred to later generations. Recombination means reshuffling of factors from one generation to the next, that is, the combination of single genes that constitutes the total genetic makeup changes from parent to offspring. The chromosomal mechanism that executes this feat was elucidated during the first and second decades of the twentieth century. This chromosomal mechanism not only explained the simple numerical rules that Mendel had observed in some ideal cases, it was soon developed to explain deviations from the rules. *The Mechanism of Mendelian Heredity* (1915) by the American zoologist and geneticist Thomas Hunt Morgan and his collaborators gave the classical presentation of the chromosome theory. Hermann Muller, who played an important role as visiting geneticist in Soviet Russia from 1933 to 1936, was a also a key person in creating this book.

According to classical genetics, hereditary variation between individuals is overwhelmingly the result of new combinations of existing genetic factors. Mutations that change the elementary factors are relatively rare. This implies that the variation that natural selection can work on to shift the average character of a population is mainly due to recombination. Natural selection sorts out those combinations that are most fit to survive. Correspondingly, artificial selection by human intervention sorts out those that fit specific human purposes.

The fortuitous character of the mutations was an essential part of classical genetics. If a mutation happens to enhance the survival of its host, it is by pure chance. But, as Darwin had explained, with enough trials even such a wasteful mechanism could produce the most wonderfully adapted living organisms. Though the variation created by recombination was most important in short-term change, evolution in the long run depended on mutations. Without mutations, evolution would come to a halt as all possible recombinations had been tried out.

The neo-Darwinian view of evolution as natural selection working on a system of stable genes appeared as a very primitive mechanical model to many biologists. They felt that mutation, recombination, and natural selection alone could not explain the wonderful adaptations and the evolutionary changes that they observed in nature. Among taxonomists, systematists, ecologists, and paleontologists there was widespread dissatisfaction with neo-Darwinism. Classical (Mendelian) genetics simply did not give natural selection enough variation to work on. Many critics felt that neo-Darwinism could not explain macroevolution, that is, the evolution of new species and of radically new organs and properties of living organisms.

The stability of the genes was continually challenged thorough the 1920s and 1930s. The idea that elementary hereditary factors can remain unchanged for hundreds and thousands of generations was criticized as metaphysical speculation. The alternative was a softer kind of genetic factor that could accommodate the inheritance of acquired characters. The main idea of so-called neo-Lamarckism is that properties that an individual organism acquires during its lifetime can become part of the genetic inheritance of its offspring. The properties that an organism develops in response to its environment, for instance, the big muscles of the blacksmith, could thus directly influence the biological heredity transferred to the next generation. This idea could be found, for instance, in Charles Darwin's *The Origin of Species*.

To uphold neo-Lamarckism as a viable scientific alternative it was necessary to demonstrate the inheritance of acquired characters. The inability to produce clear and convincing demonstrations was the weakness of neo-Lamarckism. By the middle of the twentieth century this had led to a quite marginal existence for theories of neo-Lamarckian inheritance. Nevertheless, there are still, at the end of the century, good indications that the idea has to be taken seriously. Molecular mechanisms seem to exist by which the environment can shape the genetic material in a quite fundamental way. The salient questions are how important such mechanisms have been in evolution and whether they can be exploited by man to manipulate the heredity of plants and animals (Jablonka and Lamb 1995). This was essentially what Lysenko claimed to have achieved.

GENETICS AND PLANT BREEDING

The development of large-scale production and the sale of seed in the second half of the nineteenth century stimulated a scientifically based plant breeding. The search for new methods and techniques was often based on Darwin's theory of the evolution of species. Among the three who rediscovered Mendel's laws, there was one professional plant breeder, namely, the Austrian Eric von Tschermak.

The other two—the German Carl Correns and the Dutchman Hugo de Vries—were concerned with theoretical questions on the formation of species.

Darwin's ideas about biological variation and heredity turned out to be misleading. The occasional occurrence of individuals with highly distinct and more or less heritable properties in a population was a well-known phenomenon. Plant and animal breeders had made use of such "sports," and Darwin discussed them extensively in his book *The Variation of Animals and Plants under Domestication* (1875). But his own general theory of heredity emphasized continuous rather than discontinuous change. According to Darwin's theory of pangenesis, all parts of the parent organism gave their contributions to the germ cells of the parents in the form of minute particles called pangenes. The inheritance of acquired characters was a part of his theory that became more important with successive editions of *The Origin of Species*. His theory implied continuous hereditary variation with no other limits than the viability of the plant. By selecting and reproducing a group of individuals that had a desired property in a high degree, it should be possible gradually to shift the average character of the population. This method was called mass selection, in contrast to individual selection, which used single outstanding individuals as the starting point for following generations. The latter method was also called pedigree breeding.

It turned out that plant breeders who used mass selection according to Darwinian theory often met with unexpected failure. The offspring inherited the average character of the whole parent population rather than their selected parents. The populations simply refused to be changed according to the theory of orthodox Darwinism. Typical examples turned up in the work at Svalöf in the 1880s and 1890s. This institution was soon to become the world's leading center for plant breeding. But the start was not so promising. Mass selection for properties such as stiffness of straw in barley and winter hardiness of wheat plants gave meager results or proved completely useless. Only with the introduction of pedigree breeding did Svalöf make important progress. Mass selection was not completely discarded. It was retained as an important method for more specific purposes. But picking out single plants and breeding their offspring in separation turned out to be the key to success (Roll-Hansen 1990, 1997).

Both Hugo de Vries and Wilhelm Johannsen had observed the lacking effect of mass selection. Johannsen was also well acquainted with the extensive work at Svalöf when he introduced the distinction between *genotype* and *phenotype* into genetic theory with his classical treatise of 1909, *Elemente der Exakten Erblichkeitslehre* (Elements of an Exact Theory of Heredity). According to Johannsen, the genotype represented the hereditary form that could remain unchanged for generations except for infrequent mutations. The phenotype was the concrete organism as it developed through the interaction between hereditary and environmental factors. The same genotype would produce different phenotypes under different environmental conditions. For instance, a plant could develop differently at sea level and high altitude, or in humid and dry air.

When Johannsen carried out his classical selection experiment on beans, the degree of hereditary stability took him by surprise. Though he knew from de Vries that stability was much higher than the orthodox Darwinians believed, he had expected to find that at least a little bit of the continuous variation owing to environmental influences was inherited. But there was no heritability (Johannsen 1903). This result attracted great interest and was confirmed by other researchers for a broad range of organisms, both plants and animals. Johannsen became an effective spokesman for the stability of genes, a cornerstone of classical Mendelian genetics, as well as the neo-Darwinian model for the transformation of cultivated plants. But he did not accept Mendelian genetics as the general mechanism underlying the evolution of species. Like many other pioneers of classical genetics, he thought that Mendelian inheritance could not explain macroevolution (Roll-Hansen 1978).

Johannsen's theory of *pure lines* became very important for plant breeding in the following decades, especially in Russia. It formed the theoretical basis for pedigree breeding. A pure line was defined as all descendants of a single individual. In self-fertilizing plants all individuals of a pure line would tend to have identical hereditary properties—to belong to the same genotype. But if there is no hereditary variation in a population, there will be no effect of selection. It appeared to follow from Johannsen's theory that selection or hybridization within pure lines of self-fertilizers is futile. This was one of the dogmas that Lysenko came to challenge.

Both on the level of general theory and on the level of practical plant breeding there was a lasting tension between classical Mendelian genetics and views that attributed a fundamental role to the environment. While some Mendelians had a tendency to describe phenotypic traits as determined by genes alone, more sophisticated Mendelians like Johannsen stressed that heredity and environment are equally essential in the formation of the organism. But as already mentioned, neo-Lamarckian critics believed in a direct formative effect of the environment on the genes. Only by the 1940s was a neo-Darwinian "new synthesis," which united evolution by natural selection with stable Mendelian genes, broadly accepted in biological science (Bowler 1983, 1989).

While the Mendelian theory was a product of specialized scientific research, neo-Lamarckian ideas corresponded to traditional popular views on heredity. Lamarckian ideas had intuitive appeal to practical breeders with little theoretical training in biology. Luther Burbank and Ivan Michurin are prominent examples.

In the 1930s many leading geneticists, especially in continental Europe, still doubted the general validity of classical Mendelian genetics and the closely associated chromosome theory. In their view the cell nucleus did not have the dominant role in heredity that the chromosome theorists attributed to it. Instead the cell plasma, or the cell as a whole, was seen as the seat of important hereditary factors, for instance, those that determined the differences between species (Harwood 1984, 1993; Sapp 1987). For instance, Iurii Filipchenko, holder of the first Russian chair in genetics, believed that the chromosome theory pertained only to differences within the species, while the formation of new species, macroevolution, had a different material basis and followed different laws.

Some philosophizing biologists also argued in the 1930s and 1940s that Mendelian genetics was incompatible both with the evolution of species and with the development of the individual. The concept of stable genes was merely a modern version of preformationism that assumed a blueprint or miniature model of the adult organism existed in the germ. They rejected the "mechanist" approach to biology in favor of an "organismic" view. Living organisms are essentially "wholes," and their behavior cannot be explained from the properties and relations of their chemical parts alone.[1]

Many breeders of academic as well as more practical orientation doubted the general validity of Mendelian genetics and the principles of breeding that followed from it. This was a widespread attitude both in Germany (Harwood 1997) and in England. In Germany breeders working for commercial seed firms tended to be skeptical of Mendelism, while those in government-financed academic institutions were generally positive. In England skepticism was apparently widespread among the academics as well. Plant breeding is "more art than science," wrote the professor of plant breeding at Cambridge University, England. Though genetic science had "stimulated the breeder and altered his outlook" it had not so far given him "any substantially new method" (Engledow 1931, 87). The professor questioned the stability of pure lines and genes and defended the farmers' intuition that when new varieties are propagated over generations their properties are bound to "revert" or "run back" in time. Overconfident theoreticians would limit the possibility of change to mutation or occasional crossings with other varieties, but there was evidence that forms of apparent perfect uniformity, which were homozygous according to Mendelian theory, "may retain this uniformity for only a few years when growing as large-scale field crops," argued Engledow (1931, 86). In other words, he suggested that genes were not as stable and unaffected by the external environment as classical genetics taught.

This skeptical British view of Mendelian genetics was partly a result of disillusionment with scientific plant breeding. The "wild hopes" of "immensely improved varieties" aroused in some people by the discovery of Mendel's laws in 1900 had not materialized (Engledow 1931, 90).[2] Leading breeders on the European continent, for instance, in Sweden and Germany, often had a more positive evaluation of the role of Mendelian genetics. The more positive reaction on the European continent was perhaps due to the new methods of hybridization and pedigree selection in fact being more successful there than in England. In Sweden, for instance, hybridization soon produced a number of valuable new varieties of cereals. The method of hybridization was successfully developed by Herman Nilsson-Ehle at Svalöf in close contact with Erwin Baur in Germany (Roll-Hansen 1990). It is tempting to speculate that the reason for greater success in northern continental Europe than in England was that in England the potential of recombination had largely been exhausted by nineteenth-century breeders using methods that were less theoretically precise and sophisticated but still effective in the long run (Hunter 1939; Roll-Hansen 2000).

Russia was trailing behind even more in practical application than in theoretical research. The new varieties bred by agricultural stations and seed firms made very little impact on Russian agriculture before World War I. The overwhelming majority of Russian fields continued to be sown by traditional peasant varieties. Only after the end of the civil war was the country able to resume efforts to improve the general seed material. In 1922 the Soviet government established a new system of seed management, and by 1931 about 10 percent of the country's fields were sown with selected varieties (Pangalo 1931). Technical cultures such as cotton and sugar beet were most advanced. Grain cultivation was lagging.

DEVELOPMENTAL PHYSIOLOGY: EFFECTS OF LIGHT AND TEMPERATURE

Except for greenhouse cultivation, the manipulation of plant development by light and temperature has in fact contributed little to practical methods of agriculture. However, the *possibility* of developing important practical methods attracted worldwide attention in the 1920s and 1930s. The study of how light and temperature directs plant development has also been a major theoretical topic. In particular, the long-term effects, often called "aftereffects," have drawn interest from both practitioners and theoreticians. These phenomena differ from the direct effects that light has in driving the assimilation of carbon dioxide, or that low temperature has in slowing down the metabolic processes. Aftereffects imply, for example, that specific light regimes or temperature conditions trigger a process that starts a long time after the impulse. There is a clear economic motive for the study of such aftereffects. They offer the prospect of intervention at an early stage when the plants are still easy to manipulate. Treating the seeds before they are sown is much less demanding than treating the plants when they are spread all over the fields.

Studies in the physiology of plant development were also stimulated by a growing interest in plant ecology as a theoretical subject. Toward the end of the nineteenth century, ecology was much concerned with the influence of external factors such as temperature, light, water, nutrition, and soil (Warming 1895; Schimper 1898), and ecophysiology became a special subdiscipline applying experimental methods to ecological questions.

The German plant physiologist Gustav Klebs was a pioneer in studying the succession of stages in plant development. For instance, he distinguished three main stages between the first rosette of leaves and flowering: "first, the lengthening of the axis; second, ramification; and third, the genesis of flowers." He discovered that these stages were to some extent independent and separable. Not only could he halt the plant at a certain stage and stop it from further development, he could also make it skip a stage. For instance: "When rosettes, ripe to flower, are cultivated at the end of April in a very warm soil and in strong light the flowers are developed on top of the old rosette without axis lengthening" (Klebs 1910, 552–54).

It was well known that many plants required a period of low temperature to break their hibernation. This had long been exploited in methods to induce flowering out of season. It was also tried on winter grains. Usually winter grains are sown in the late autumn and flower the following summer. If sown in spring, they will be slow in flowering or remain in the vegetative state throughout their first growth season. But a period of low temperature, after moistening and slight germination, can produce normal flowering and harvest. Attempts to apply this phenomenon to improve practical grain growing had not proved profitable, however (Liubimenko 1933, 4; Zirkle 1949, 27).

The German botanist Gustav Gassner was a pioneer in studying the cold requirement of cereals. In the subtropical climate of Uruguay he observed how winter cereals sown in spring were unable to spike even with a very long growth period, while autumn sowings would spike readily in spring. However, he found that if the seed was germinated slightly at temperatures not higher than 6 to 10 degrees centigrade, spring or even summer sowings would spike rapidly. After returning to Germany he continued this work with the help of refrigerated rooms. At regular intervals from January to July seed were germinated at temperatures ranging from 1 to 2 degrees centigrade upward. After three weeks of this regime they were planted in the field. It turned out that with a germinating temperature of 5 to 6 degrees or higher, spiking was poor when planting occurred after the end of March for a certain variety of winter rye, and after the end of April for a certain winter wheat. Gassner concluded that flowering in obligate winter annual plants is necessarily dependent on a period of low temperature at an early stage in their development (Gassner 1918, 431). His interest in this phenomenon was by no means unique. It was studied at the same time by many other researchers, for instance, in Russia (Murinov 1913, 1914).

Gassner's 1918 paper contains an intricate discussion of the interaction between inherited "norms of reaction" and external environmental conditions. He had a clear conception of the distinction between a phenotypic character such as time of flowering and the underlying inherited norm of reaction. He was also convinced that the old Lamarckian idea about a gradual hereditary adaptation to different environmental conditions had been disproved. Gassner's aim was to find the physiological mechanism that is triggered by the low temperature and thus links the norm of reaction (genotype) to a certain type of behavior (phenotype) of the plant. His paper thus demonstrates the close connection between questions of physiology and genetics at this time.

The discovery of the photoperiodic effect is usually attributed to a 1920 paper by two Americans (Garner and Allard 1920). Some plants need short days for full development. Only with nights longer than a certain minimum will they set flowers. Long-day plants, on the other hand, demand more than a minimum period of daylight. Since the length of day varies cyclically through the year, and more so the farther one is removed from the equator, this provides a mechanism for the regulation of plant development through the year. Photoperiodism and aftereffects of low temperature (vernalization) are the two main mechanisms for keeping plants in tune with annual climatic cycles. In the 1920s and 1930s the study of these two were among the hottest topics in plant physiology.

Alongside this experimental tradition in academic ecophysiology there existed a popular movement called phenology, which was devoted to studies of the annual life cycles of plants and climate. The year-by-year registration of key dates by the phenologists, for instance, the onset of flowering of a certain plant, was aimed at discovering regularities in nature. This was an old botanical interest counting Carolus Linnaeus among its fathers. The name "phenology" was introduced when a broad movement developed in the mid-nineteenth century. A pioneer and activist was the versatile Belgian statistician and demographer Adolphe Quetelet. Revival of phenology in the 1920s and 1930s was motivated in part by agricultural interests.

The *Phänologische Reichsdienst* (National Phenological Service) in Germany was organized just after the First World War. One of its main tasks was to observe the development of winter cereals. By 1939 this organization had a large network of observation stations and ten thousand collaborators. In 1924 the phenological committee of the Royal Meteorological Society in London proposed a system of cooperation for all of Europe. This system was modified by the *Internationale Kommission für Landwirtschaftliche Meteorologie* (International Commission for Agricultural Meteorology) in Munich in 1932 with special consideration of agricultural needs. In 1928 there existed a system of more than 2,000 observation posts on the European continent and 350 on the British isles. But international cooperation came to a halt as World War II started in 1939–1940 (Printz 1965). The movement has not recovered its old vitality since then.

Phenology was a big movement in popular science that has

now been largely forgotten. Its simple descriptive and statistical character did not appeal to the following generations of experimentally oriented modern biologists. But it represented important biological interests of its time. Phenology was easy to grasp for a person with little scientific training, and it was an important ingredient in worldwide agricultural science in the period between the two world wars. Lysenko's early work on plant development in part grew out of this tradition.

Phenology was widely present in the Soviet Union, though it was apparently never institutionalized as a special botanical subdiscipline. A 1937 Moscow meeting of phenologists agreed on the urgent need for a nationwide phenological service modeled on the meteorological service. They also wanted a "scientific-methodological phenological center" under the Academy of Sciences. The introductory speaker claimed that the network of observers had grown substantially since the revolution and presently consisted of twenty-five hundred to three thousand phenologists (Pechnikova 1937).

THE IMPERIAL LEGACY OF SOVIET AGRICULTURAL RESEARCH

Many European countries established new institutions for agricultural education and research during the mid-nineteenth century. In Russia the Petrovka Academy for Agriculture and Forestry was opened at Moscow in 1865, following the abolition of serfdom in 1861. The Petrovka, later renamed Timiriazev Agricultural Academy, has remained the central scientific institution of Russian agriculture up to the present.

The Petrovka was established in the spirit of nineteenth-century liberal enlightenment. At first, there were no formal requirements for admission and no obligatory exams. In 1869 the school wanted to admit female students but the ministry refused. Soon a reputation for being a "revolutionary hotbed" made the government establish tight control on student activities. But the Petrovka remained a center of liberal political opposition.[3]

The botanist Kliment Timiriazev (1843–1920) started teaching at the Petrovka in 1870. He was known as a main proponent of Darwinism in Russia, and in old age he became a Bolshevik hero because he was one of the few prominent scientists who wholeheartedly supported the new regime. In 1877 Timiriazev was chosen for a new chair in plant physiology and anatomy at Moscow University, but he continued teaching at the Petrovka until he was dismissed for his radical political views in 1894. Like many Darwinians of his generation, Timiriazev was skeptical of Mendelian genetics. His writings later became an arsenal for the Lysenkoists in their struggle with classical genetics (Gaissinovich 1985). Timiriazev died in 1920, before he had a chance to take part in the struggle, and in 1923 the Petrovka was renamed in his honor.[4]

After a period of political repression and stagnation, the Petrovka was reorganized in 1894. The government attempted to reactivate the institution without its liberal political tradition. Officially it was renamed *Moskovskii sel'skokhoziaistvennyi institut* (Moscow Agricultural Institute), but in daily parlance the old colloquial name, Petrovskaia or Petrovka, was still used. This reform brought a new era of expansion with rapidly increasing student numbers as well as the expansion and diversification of scientific research.

Nitrogen fertilization and nitrogen metabolism was a hot topic in agricultural research at the end of the nineteenth century. Russian soil scientists, microbiologists, and botanists made important contributions to international science. At the Petrovka studies of nitrogen in plant nutrition was pursued by Dimitrii Prianishnikov (1865–1948), who took over the chair of plant cultivation (*chastnoe zemledelie*) in 1895. He developed methods for nitrogen fertilization and became an advocate for the use of chemical fertilizers and the development of a chemical industry to produce them. In the 1920s he took an optimistic view of agricultural productivity, believing that by the time the population of the Soviet Union had doubled, agricultural production could be increased six- or sevenfold. He was active in working out the five-year plans for agriculture (*Moskovskaia sel. Akad.* 1965, 396–97). Prianishnikov showed an extraordinary ability to cooperate, at arm's length, with Stalin's regime and still keep his intellectual integrity and moral balance. He voiced candid criticism of official views and decisions that he found mistaken, and he repeatedly intervened on behalf of other scientists who were suppressed. But apparently he kept the confidence of the government. Until his death in 1948 Prianishnikov retained his leading position as an authority and manager in agricultural science.

Vasilii Vil'iams (1863–1939) was called to the Petrovka as professor of soil management (*pochvovedenie*) in 1894. He developed a vitalist view of soil processes, often expressed through analogies between the soil and the living organism (*Moskovskaia sel. akad.* 1965, 403). In the 1920s and 1930s there was intense rivalry between him and Prianishnikov. Vil'iams worked out a system of plant rotation (*travopol'e*) with extensive use of grassland including legumes. This system was supposed to stimulate the nutritional regeneration of the soil and thus reduce the need for artificial fertilizers. Large investments in chemical factories could be saved. Politically Vil'iams was an opportunist who cooperated closely first with the tsarist government and later with the Bolsheviks. He became an active supporter of Lysenko. The term "agrobiology" covered Vil'iams's as well as Lysenko's type of agricultural science.

The systematic development of a network of agricultural experimental fields and stations throughout the Russian empire started in the mid-1890s. The first stations were established in cooperation among the central government, *zemstvos* (local government), and big landowners. This work was sped up after the almost-revolution in 1905. An all-Russian conference on the organization of agricultural experimental work in 1908 laid down the principles of a program that influenced developments well into the Soviet period. This program envisaged a regional organization with one central station directing the work in each region. In 1912 it was officially accepted with revisions and specification in the form of a law (Elina 2002). By 1913 there existed altogether 168 experimental stations in the whole of Russia. At the head of this network were five large stations in the five regions of "Middle-Russia," in Moscow, Kharkov, Kiev, Ekaterinoslav (now Dnepropetrovsk), and Saratov (Stebutt 1913, 39).

Simultaneously with the renovation of the Petrovka in 1894, a new Bureau of Applied Botany was founded in St. Petersburg. Its task was to make surveys and investigations of cultivated plants throughout the European and Asiatic parts of the Russian empire, to assist the introduction of new plants and cultures, to collect relevant information from other countries, and to make botanical studies of Russian cultivated plants and weeds. In the course of his investigations of barley, Robert Regel worked out a system of classification and assessment of varieties that was later taken over and developed by Nikolai Vavilov. After Regel became head of the bureau in 1905, the activity grew rapidly. Between 1907 and 1912 its budget increased forty times (Lassan 1997). By the start of World War I the bureau was a large institution with branches in many parts of the country (Pisarev 1929). When Regel died in 1920, Nikolai Vavilov took over and developed the bureau into the famous All-Union Institute of Plant Industry, VIR (*Vsesoiuznyi Institut Rastenievodstva*).

Modern scientific plant breeding in Russia was developed within this network of experimental fields and stations under the leadership of central state institutions, primarily the Bureau of Applied Botany and the Petrovka. The first scientific breeding work was undertaken at the Petrovka by Dionisy Rudzinsky from 1902 onward. In 1906 a special plant breeding station was established here. It was officially opened in 1908 as a central teaching institution (Stebutt 1913, 41; Elina 1997a).

The development of plant breeding and other agricultural research after the 1917 revolution was to a large extent built on plans from the old regime. The general plan of 1908 was reworked in 1919. Compared to the United States, agricultural research in imperial Russia had a tradition of strong central planning (Pisarev 1929, 175). Its structure had more in common with the system of state-supported cooperative organizations that was developed in the Scandinavian countries.

A Bolshevik reorganization in 1923 made the agricultural research system even more centralized. A large number of new experimental stations had sprung up in the years immediately after World War I (Elina 1997b). Partly, this was a way for scientists to survive in a society with chronic lack of food and other basic means of subsistence. There were many different possibilities of patronage and public support in the pluralistic early Soviet system. But these new stations were often poorly equipped and precariously funded. A government commission on the organization of agricultural research advised extensive closure or transfer of funding responsibility to local authorities (Elina 1997b).

The recommendations of the 1923 commission substantially reduced the autonomy and influence of working scientists in favor of bureaucrats. The relatively flat structure of the existing system of the agricultural research network was transformed into a pyramidal hierarchy. Central research institutes with a network of local branches were to take care of the more basic research. The regional experimental stations were downgraded to more local and practical tasks. It was a typical example of implementing the so-called linear view of technical innovation: innovation starts from new theoretical knowledge, which gives rise to technical inventions, which are then set to solve practical economic tasks. One effect of the reorganization was increased separation of basic science from practical work. First to get the status of central national institute was the Bureau of Applied Botany, now called the All-Russian Institute of Applied Botany and New Crops. Two of the three scientific members of the commission[5] protested strongly but with little effect (Elina 1997b).

SOVIET PLANT PHYSIOLOGY AROUND 1930

The first Russian chair in plant physiology was established at the University of St. Petersburg in 1867. In 1890 the Academy of Sciences, then located in the same city, founded a "cabinet" for research in the same subject. St. Petersburg remained the dominant center of Russian plant physiology, as it was also the scientific capital of the country, until the transfer of the Academy of Sciences to Moscow in the mid-1930s.

The effect of low temperature on winter annual grains was studied at the Petrovka under the supervision of Prianishnikov. In contradiction to Gassner, it was concluded that neither low temperature nor rest (hibernation) was strictly necessary for winter annual cereals to spike (Murinov 1913, 1914). But the plant material, mainly rye, was genetically nonhomogeneous, and the significance of the experiments was problematic.

In the 1920s the main Soviet institutions for research on plant development and ecophysiology were the laboratory of Vladimir Liubimenko at the Leningrad botanical garden and the physiological laboratory of VIR headed by Nikolai Maksimov. Only in 1932 was developmental physiology taken up as a major research topic by the Academy of Sciences' Laboratory of Plant Physiology and Biochemistry.

Vladimir Liubimenko (1873-1937) was the pioneer of Russian ecophysiology. After graduating from the Institute of Forestry at St. Petersburg he continued his studies abroad. Back in Russia he organized a physiological laboratory in the Nikitskii botanical garden in the Crimea, and from 1914 he worked in the St. Petersburg botanical garden. His central interest was plant pigments, photosynthesis, and other reactions of plants to light (Maksimov 1947, 243ff). With these interests it was natural for Liubimenko and his collaborators to study photoperiodism when this phenomenon was discovered in 1920.

Nikolai Maksimov (1880-1952) started his research career with experimental and theoretical studies of the mechanisms of winter hardiness and drought resistance. In 1922 he was invited by Nikolai Vavilov to organize the laboratory of plant physiology at the Institute for Applied Botany (later VIR). The laboratory was located in Detskoe Selo just outside of Leningrad in conjunction with the central station for plant breeding. It quickly became the national center for agricultural plant physiology. But after Maksimov was arrested in 1933 and later banished to Saratov its importance declined.

Maksimov investigated contradicting results in earlier experiments on cold requirement in winter cereals. He confirmed that low temperature was not an absolute requirement.[6] As we shall see, Lysenko later sided with Gassner, claiming a period of low temperature to be necessary.

Another branch of developmental plant physiology emerging in the 1920s was the study of hormones. The first plant growth hormone was discovered in 1926, simultaneously in the Netherlands and in the Ukraine, in Kiev by Nikolai Kholodnyi (Manoilenko 1969).

In 1932 the Academy of Sciences established a special laboratory of plant physiology and biochemistry under Andrei Richter (1871–1947). At this point Lysenko was already famous for his work on vernalization. The publicity that Lysenko's work created was important for the quick expansion of Richter's laboratory. In 1934 it moved from Leningrad to Moscow and reconstituted as the academy's Institute for Plant Physiology. The current political atmosphere demanded practical relevance, and experimental laboratory studies of plant development became an important part of the work. One group, headed by Richter himself, worked on vernalization, focusing on the biochemical processes underlying the phenomenon. Another group, led by Mikhail Khristoforovich Chailakhian (1902–1991), studied photoperiodicity. Through ingenious experiments Chailakhian demonstrated how the leaves acted as receptors of the light stimulus, which was then transmitted to the growing point where the flower buds were formed. He proposed the existence of a flowering hormone, "florigen," as responsible for these effects (Kholodny 1939, 225–29).

Lysenko was in general opposed to reductionist chemical explanations of plant development. During his days of power, Chailakhian and Kholodnyi came to suffer for their support of hormone theories of plant development. Similarly, other plant physiologists came into conflict with the more or less obscure holistic views of Lysenko and his followers. The repression was less severe than in genetics, however. Most plant physiologists were able to continue research within their field throughout the Lysenko period with some adaptations and concessions to agrobiological ideas. But the cost in terms of scientific integrity and personal strain was often high, and some did not survive.

SOVIET PLANT BREEDING IN THE 1920s

On the initiative of Vavilov a Central Experimental Station for Genetics and Plant Breeding was founded in 1921 in Detskoe Selo, just outside of Leningrad. In 1925 the leader of this institution,

Viktor E. Pisarev (1882–1972), published a survey of Soviet plant breeding. He underlined the continuity and dependence on prerevolutionary efforts at the same time as he proudly presented achievements under the new regime. While a similar survey in 1913 was mainly concerned with organization and programs for future work, we can now already speak of "important scientific and practical results of Russian plant breeding," he concluded (Pisarev 1925, 221).

Pisarev described important results from various regions of the Soviet Union. There was, for instance, very promising work on distant hybridization going on at Saratov on the lower Volga. From naturally occurring hybrids between wheat and rye one was hoping to breed wheat with the winter hardiness of rye. The lack of hardy winter wheats was a chronic problem, so this "would be the solution of a very important task" (Pisarev 1925, 236). Among the practical achievements of the Sibirean stations of Tomsk and Irkutsk were a number of early ripening spring wheats. Some of the wheats from Irkutsk had proved to be the only ones suitable for growing in central Alaska, wrote Pisarev with some pride (244). He had himself been head of that station for a while. Both in Tomsk and in Irkutsk the main method was pedigree selection without any conscious application of hybridization. This was the quick way of making use of the variation present in the indigenous populations.

Pisarev gave a detailed description of the selection work on sugar beet in the Ukraine. Before the revolution a thriving Ukrainian seed business supplied seed of sugar beet as far away as the United States, England, and France. This private seed production was destroyed in the civil war. But under the state-owned Sugar Trust a new network of plant-breeding stations was formed. By the end of 1923 the Sugar Trust had fourteen such stations working with vegetables and grain as well as sugar beet. Among these the Belaia Tserkov station, where Lysenko worked in the early 1920s, was listed as doing mass selection on sugar beet and individual selection on wheat, oats, and vegetables (Pisarev 1925, 249). In other stations sugar beet was subjected to both mass and individual selection, and according to Pisarev the leaders of sugar beet breeding were becoming convinced of the "the uselessness of such mass selection" (248). In a second survey of 1929 Pisarev described the Sugar Trust as the "biggest organization for plant breeding not only in the Ukraine, but in the whole Union" (Pisarev 1929, 217).

NIKOLAI VAVILOV—ENTREPRENEUR OF SCIENCE

Nikolai Ivanovich Vavilov (1887–1943) was the son of a prosperous Moscow merchant who had worked his way up from peasantry. His thorough education had a modern and practical rather than a traditional academic orientation. After finishing a commercial secondary school he entered the Petrovka. Upon graduation he was trained as a specialist in plant breeding. In 1913–1914 he traveled and studied in western Europe, making his most important contacts in England. He studied in Cambridge with the plant breeder Prof. R. H. Biffen. But the greatest impression was left by William Bateson of the John Innes Institute of Horticulture. Vavilov later referred to Bateson as his main teacher in genetics.

In the summer of 1917, after the tsar had been overthrown in February, Vavilov became professor of plant breeding in Saratov. His father left Moscow after the Bolshevik coup in October 1917 and participated in attempts to overturn the government, but the two sons, the biologist Nikolai and the physicist Sergei, stayed and worked loyally for the new regime. Sergei Vavilov in due time became president of the Academy of Sciences and was destined to preside over the suppression of genetics in 1948.[7]

Nikolai Vavilov was full of enthusiasm for science as a progressive social force. He was a practically oriented scientist, an organizer, and also a public spokesman for science. Vavilov's optimistic and liberal view of science had much in common with that of the British science writer and editor Richard A. Gregory. A Russian translation of his book *Discovery, or the Spirit and Service of Science* was edited by Nikolai Vavilov and translated by his first wife (Gregory 1923). In the 1920s and 1930s Gregory played a central role in the British scientific establishment. For many years he was editor of the journal *Nature*. He shared with the young generation of radical British natural scientists such as

The Situation in Agricultural Plant Science | 41

John Desmond Bernal, J. B. S. Haldane, and Joseph Needham the wish that science should become more useful and more conscious of its social responsibility. But he did not share their sympathy for Marxism. Gregory worried that science had often been used to "convert beautiful countryside into grimy centers of industrialism, and to construct weapons of death of such a diabolical character that civilized man ought to hang his head in shame of their use." However, he thought the responsibility "does not lie with scientific research, but with statesmen and democracy" (Gregory 1921, 448).

Nikolai Vavilov is often depicted as the leading geneticist of the Soviet Union. His most important scientific contributions were, however, in plant geography and systematics rather than in genetics or other experimental disciplines. In particular, he worked on the systematics of cultivated plants and their varieties. Vavilov's paper "The Law of Homologous Series in Variation" (1922) describes how the same series of characters are found in related species, for instance, in wheat and rye. He argued that this parallelism was a result of similar underlying genetic dispositions. In 1930 he lectured to the International Botanical Congress in Cambridge on "The Linnean Species as a System," applying his idea of homologous series to intraspecies systematics (Vavilov 1931c). These two papers laid the theoretical basis for the "World Collection" of cultivated plants that Vavilov built at the All-Union Institute of Plant Industry (VIR) in Leningrad.

In retrospect the World Collection appears as Vavilov's most important scientific achievement. It was the world's first gene bank. Material was collected by Vavilov and his collaborators on numerous expeditions both within the Soviet Union and abroad. Its purpose was to serve the breeding of new plants for Soviet agriculture. The development, maintenance, and practical application of this collection was the main project of Vavilov and VIR throughout the 1920s and 1930s.

Vavilov's most significant scientific contribution is probably his theory about the origin of cultivated plants. According to this theory most cultivated plants came from a number of limited geographical areas, which corresponded to the areas with the most variability at present. Vavilov's theory was soon subjected to severe criticism. In response Vavilov blurred its features. Both the number of centers and their geographical extension was increased. Nevertheless, Vavilov's 1926 book, *The Centers of Origin of Cultivated Plants*, remains a classic in plant geography.[8]

It was as an organizer and a teacher that Vavilov made his unique contribution to Russian agricultural science. After he was called to Leningrad in 1920 to head the Bureau of Applied Botany, Vavilov soon became the leading entrepreneur in Soviet agricultural science. When the Lenin Academy of Agricultural Science (VASKhNIL) was established in early 1929, Vavilov was the natural choice as president of the new organization. In the decade from 1923 to 1934 he was the key person in the construction of an enormous centralized research system with no parallel in any other country.[9] With the reorganiztion of the VASKhNIL in 1935 Vavilov had to take much criticism for the lacking quality and efficiency of the system, but he replied with pride that "the greatest positive event of the preceding period has been the organization of a uniform network of scientific research institutions which for the first time embraces all important branches of agriculture" (Vavilov 1935a, 17).

Saratov was already a main center for agricultural research when Vavilov arrived in 1917.[10] In 1913 the station for plant breeding (*selektsiia*) in Saratov was described by its director as "one of the largest in Russia and perhaps after Svalöf one of the largest in Europe" (Stebutt 1913, 53). Two other scientists who were in Saratov at this time came to play an important role in the politics of agricultural research in the 1930s, namely, the plant breeder and geneticist Georgii Meister and the agronomist and soil scientist Nikolai Tulaikov. Both grew up under much more modest social conditions than Nikolai Vavilov, and in contrast to him they became members of the Communist Party, both in 1930 during Stalin's great break. They were also more than ten years older than Vavilov and had a career of practical work behind them before World War I started in 1914. Both were arrested and executed in the summer of 1938.

Georgii Karlovich Meister (1873–1943) was the son of an artisan. He graduated from the Novo-Aleksandrinskii Sel'skokhoziaistvenny Institut in 1897.[11] Radical political activities were an obstacle to his early career. In 1914 Meister became head of the plant-breeding station in Saratov but was drafted and did not function in this post until 1918. His son was killed in the civil war on the red side. From 1920 Meister was also responsible for the teaching of genetics and selection at the Saratov agricultural institute.[12]

Meister took Marxist ideology to heart. Besides being a successful practical breeder with a number of important grain varieties to his credit, Meister had a taste for speculative theorizing and philosophy. In the early 1930s he invested much energy in working out a truly dialectical and materialist theory of genetics and plant breeding. His somewhat naive faith in abstract thinking is illustrated by a letter of December 1934 to the American geneticist Hermann Muller, then working in Moscow. Meister explained that he was "working at present on a theoretical investigation: the analysis of the fundamental conceptions of genetics from the standpoint of materialistic dialectics." The immutability or changeability of the gene was a crucial issue. Since Muller was world famous for his work on mutation by x-rays, Meister asked him for a "definition of heredity" that could help clear up the confusion.[13]

In an authoritative textbook of plant selection Meister (1936) wrote the introductory "Methodological Analysis of Basic Concepts in Genetics." He praised the achievements of classical genetics, the distinction between genotype and phenotype, the rejection of the inheritance of acquired characters, and the chromosome theory. But his effort to give everything a dialectical and dynamic interpretation produced lengthy and sophistic theoretical analyses of dubious value to students.

A favorite idea of Meister's was distant hybridization in grain breeding. When Vavilov expressed a skeptical attitude toward his approach, Meister wrote a long defense in response (Tiumiakov 1967; Meister 1927). The disagreement was largely one of strategy. Vavilov did not deny the long-term fruitfulness of distant crossings, but hybridization of more closely related forms would bring quicker results, which were much needed.

Nikolai Tulaikov (1875–1938) grew up in a peasant family. As a student he worked in Prianishnikov's laboratory of plant nutrition. He spent two years studying in America and western Europe in 1908–1910. In 1916 he was named a member of the scientific committee of the Ministry of Agriculture and two years later president of the same committee. In 1932 he was elected a member of the Academy of Sciences in the technological section. Besides Vavilov, Tulaikov was perhaps the most important manager and entrepreneur of agricultural science in the 1920s and early 1930s. At the sixteenth party congress in 1929 the two of them gave very optimistic speeches on the possibilities that collectivization opened for the application of science to agriculture.

Among Tulaikov's specialties was soil treatment, methods of plowing, and the like in the southeast. He became a sharp critic of Vil'iams's scheme of *travopol'e*, which he claimed would have particularly negative effects in the dry and unpredictable climate of the lower Volga region. This controversy reached a high point at the All-Union conference on drought in 1931, and Tulaikov continued as a persistent critic until his arrest in 1938.[14]

In accordance with its ideology, the Soviet government promoted popular participation in science.[15] Dedicated amateurs enjoyed broad sympathy and were extolled and admired at great length in the mass media. One man reached saintly status and became the patron saint of Soviet agricultural science. Even the geneticists paid homage to him, with some reluctance.

Ivan Michurin (1855–1935) was a true self-educated amateur. He had no formal scientific training. His impoverished landed-gentry family could not even afford to put him through secondary school. But he managed to make a modest living putting all his enthusiasm into the running of a nursery on the family property. Attempts to get government support for his experiments in the breeding of fruit trees had no success before the 1917 revolution came.[16]

Vavilov first met Michurin in 1920. His belief in the value of Michurin's work was confirmed during a visit to the United States in 1921 when he learned that it was also known to American and Canadian specialists. Vavilov was also inspired by a visit to Michurin's American counterpart, Luther Burbank (Esakov and

Levina 1987, 56). It was on the recommendation of Vavilov that the Soviet government started to give Michurin substantial support for his breeding work. With the support of Vavilov he was also elected to the Academy of Sciences in 1934.

However, the later Soviet public picture of Michurin as a breeder who had produced a wealth of valuable varieties and pioneered fruit breeding in Russia was not realistic. When the fruit specialists of VIR in 1931 made a general survey of varieties worth recommending to growers, only one of Michurin's varieties was on the preliminary list. Even after reprimands from the press and the Central Control Commission of the party, the specialists gave only minor concessions (Joravsky 1970, 70-72).

Vavilov was no doubt well aware of Michurin's weaknesses as a scientist and the limited value of his new varieties. But rather than stating clearly the shortcomings, he glossed over them, praising Michurin in a way that confirmed the heroic picture of the mass media. Vavilov's public statements could easily be taken to confirm the genuine scientific and commercial value of Michurin's results. Characteristic of Vavilov's low-key criticism of the nonscientific "artisans" in agricultural innovation was his introduction to a popular biography of Luther Burbank. Here he described Burbank as an artist whose intuitions not seldom ran "counter to the established results of contemporary genetics." When the American plant breeder and geneticist George Shull was commissioned by the Carnegie Foundation to give a scientific description and evaluation of Burbank's work, he found that there existed "no strictly objective data" and had decided to give up the mission, tells Vavilov. He nevertheless praised the practical value of such "artist-selectionists" as Burbank who "intuitively find the correct way to the solution of problems of selection." But he judged it difficult to learn anything from their writings (Vavilov 1930b, 13-14).

Vavilov was particularly attracted by two features of Michurin's work, according to an American fruit-breeding expert of Russian extraction. First, Michurin was ahead of most plant breeders in hybridizing different species and in utilizing geographically foreign stocks. Second, he had made some progress in his attempts to breed varieties that could grow in northern regions that so far "nobody had considered for raising fruit" (Lang 1956). A romantic vision of settling and cultivating formerly uninhabitable dry or cold regions was an important element of Soviet policy of this period.

By the early 1930s the old Michurin had become enormously popular as a plant wizard. Lysenko tried to approach him with the aim of collaboration, but without success. Only after Michurin's death in 1935 was Lysenko able to effectively appropriate the name and market his own agrobiology as "Michurinism."

ANTI-MENDELISM IN SOVIET BIOLOGY

Established Russian biologists tended to sympathize with neo-Lamarckism and consider classical genetics as excessively mechanistic and insufficient to explain the evolution of species. Prominent examples were the plant physiologist Vladimir Liubimenko and the botanists Vladimir Komarov and Boris Keller. Born in 1873, 1874, and 1869, respectively, they were in their mid- to late forties at the time of the October revolution.

As the nestor of Russian plant physiology Liubimenko was an important guiding influence and gatekeeper of the discipline when Lysenko started his career. His *Course in General Botany* (1923) included a clear account of classical genetics. He accepted natural selection as an important force in evolution but found it insufficient for a general explanation of evolution. He also agreed that through Johannsen's experiments the scope for Lamarckian inheritance of acquired characters had been severely limited (Liubimenko 1923, 824). But it had not been eliminated.

Liubimenko speculated that so-called induced modifications sustained by environmental pressure for many generations could eventually turn into mutations (882). The same idea is found in his textbook for high school (Liubimenko 1924). He argued for "the presence in living matter of a special active and creative principle (*nachala*), which is the immediate source of new forms, new species" (Liubimenko 1923, 885). Darwinian chance variation and natural selection is far from sufficient to explain "that plan-like historical evolution of the plant organism, which we can observe in reality" (Liubimenko 1923, 889).[17] In Liubimenko's explanation of evolution there was room for orthogenetic as well as neo-Lamarckian factors in addition to Darwinian natural selection. Toward the end of his life, around 1935, Liubimenko apparently turned away from neo-Lamarckism (Manoilenko 1996, 63–65).

Vladimir Komarov (1869–1945) was a taxonomist, an ecologist, and a specialist on the floras of Kamchatka and Manchuria. From 1899 he was on the scientific staff of the St. Petersburg botanical garden. As vice president of the Academy of Sciences from 1930 to 1936 and president from 1936 to 1945 he held central positions of academic power during the early phase of Lysenkoism.

In the introduction to a new Russian edition of Darwin's *Origin of Species*, published in 1937, Komarov criticized geneticists' sharp distinction between genotype and phenotype. He also wrote that Lysenko's experiments on cross-fertilization within varieties were "carried out in accordance with Darwin and are of colossal interest to us" (Komarov 1945, 119–20). In a monograph on the nature of species published in 1944, Komarov characterized neo-Darwinism as inadequate for a general theory of evolution, but admitted it as an interesting working hypothesis. He indicated two main disagreements with the geneticists. First, botanists see variability as a direct response to environmental factors. It represents an adaptation to the environment. Geneticists hold that variability depends only on internal processes, according to Komarov. Second, botanists conceive species as "movement," as a mere stage in a process of evolution, while geneticists have an unhistorical view (Komarov 1944, 4).

Are acquired characters inheritable? This was not the right question, according to Komarov. Instead we ought to ask: Under what circumstances is adaptive variability inherited? And his answer was: Only when it affects the formation of the sex cells or the zygote. Recent investigations have demonstrated that this is possible, and thus "the chromosome theory has lost its unpleasant taste" (Komarov 1944, 105). Komarov, like Liubimenko, speculated that small and directed mutations may result from the enduring pressure of environmental factors (167ff).

Boris A. Keller (1874–1945) went further than Liubimenko and Komarov in his promotion of neo-Lamarckian views and became a direct supporter of Lysenko. Keller was a botanical ecologist. He became a member of the Communist Party in 1930 and a full member of the Academy of Sciences as well as director of the academy's Botanical Institute in 1931.[18] In 1935 he also became an academician of the Lenin Academy of Agricultural Science (VASKhNIL).

Keller actively supported Lysenko when the genetics controversy began in 1935–1936. However politically opportune this support may have been, it was also quite in accordance with Keller's own views on how the environment could influence heredity. In 1933 the press for agricultural literature had published a popular introduction to genetics by Keller. A reviewer criticized the book for its Lamarckian tendencies, or, more precisely, for its claim that under certain circumstances modifications owing to the environment could turn into mutations (Petrov 1933, 44). In a reply Keller confirmed this interpretation. He explained that he did not accept a direct inheritability of modifications. So far he completely agreed with Mendelian genetics. But he tried to establish their "possibility of changing into mutations." I am "deeply convinced" that this connection exists and needs to be studied, wrote Keller (1933, 55–56). Lysenko's program of breeding new plants through "education" by the environment was built on a similar idea, and it was natural for Keller to support Lysenko's general ideas. In fact, Keller had already presented his view of modifications changing into mutations more than a decade earlier and argued that this phenomenon should be exploited in plant breeding (Keller 1920).

The openness toward neo-Lamarckian ideas on heredity that Liubimenko, Komarov, and Keller showed was typical of the larger biological community of the 1920s and 1930s. They acknowledged that important progress had recently been made by so-called Mendelian genetics but doubted the validity of bold generalizations made by enthusiastic young geneticists. Their attitude accorded well with traditional scientific skepticism and rational conservative restraint.

Liubimenko, Komarov, and Keller had a traditional academic background. They were trained as botanists and therefore were less oriented toward practical tasks than agricultural scientists such as Prianishnikov, Meister, Vavilov, and Tulaikov. But all three were sympathetic to the social and economic aims of the new Soviet regime and sincerely wanted their science to help the progress of the country. Keller in particular pressed hard on the practical usefulness of botany. The early issues of the new scientific journal *Sovietskaia Botanika* (Soviet Botany), which started under his editorship in 1933, are charmingly enthusiastic in their approach to practical agricultural problems. But no doubt it was clear also for Keller that it was through advances in theoretical understanding that botany was most valuable to the progress of the country. Academic botany should increase its usefulness not by becoming a practical activity, but by communicating better with those engaged in practical work in agriculture, industry, and so on. By taking their problems seriously and listening attentively to their experiences, botany itself would make more rapid progress.

A different and thoroughly pragmatic attitude toward science, praising practical men and disregarding the theoretical and academic science, is found in *The New Earth*, written by an American popular-science writer (Harwood 1906) and published in Russian in 1909. The book describes the revolution that American agriculture was going through and how this was bringing radical change to the whole of society. The driving force was science: not the science of the old academics, but that of "another class of scientists" with technological creativity and a keen interest in practical affairs. The model scientist of this book was Luther Burbank, "the greatest of all plant breeders." He united theory and practice as "the most practical of all men" and "preeminently the man of science . . . continually adding to the total of human knowledge" (Harwood 1906, 117–23). The Russian edition (1909) was prefaced by Timiriazev and is reported to have highly impressed Lenin's thinking about agricultural science in the early 1920s.[19] *The New Earth* represented a populist approach to science that was to become much more influential in the Soviet Union than in the United States. The book contained little precise information, either about the theoretical ideas that guided Burbank's work or about the contributions of traditional science to agriculture. It expressed a superficial and blue-eyed pragmatic enthusiasm for science, an easy prey to wishful thinking.

Several circumstances made the problem of scientific quality control particularly acute for the young Soviet regime. A rapid expansion of the research system, combined with political suspicion toward the specialists, made the overall political steering difficult. As in Europe and North America at the beginning of the twenty-first century, strengthening central political control appeared as a way to secure that government resources were actually used for their intended purpose. Marxist theory, with its principles about primacy of practice over theoretical knowledge, supported such a move and indicated ways of implementing it. The first stage of bureaucratization in Soviet agricultural research was marked by the 1923 reform. Protests from leading scientists that this would impair the quality of the science were not much heeded. In addition to the external pressures came attitudes and strategies developed by the scientists themselves. Under the leadership of Nikolai Vavilov, agricultural science stimulated and used the Michurin myth to attract public enthusiasm and funds for their research. The utopianism and the scientific vagueness of this legacy became a problem when Lysenko managed to grab the mantle of Michurinism.

NOTES

1. See Ruse (1975) and Roll-Hansen (1984) on the holism of E. S. Russell and J. H. Woodger. An English defender of Lysenko's genetics, A. Morton, in his book *Soviet Genetics* (1952) referred to Woodger's *Principles of Biology* (1929) for support. Morton's book was translated into Russian.

2. This British pessimism is described by Palladino (1994).

3. For detailed information about the Petrovka (or Petrovkskaia) Academy I depend largely on the book published at its centenary: *Moskovskaia sel'skokhoziaistvennaia akademiia K. A. Timiriiazeva* (1965).

4. The full official name was *Moskovskaia sel'skokhoziaistvennaia akademiia imeni K. A. Timiriazeva*.

5. A. G. Doiarenko and N. M. Tulaikov.

6. Two series of seed were sown, one in a heated greenhouse (ca. 15

degrees centigrade) and the other in an unheated greenhouse (ca. 5 degrees centigrade), from late December to April. Between May 15 and 20 the plants were transferred to the field. Normal spiking was obtained for winter wheat of both series sown up to March 3, though development was speeded up by low temperature. But by the sowing of April 16 there was a marked difference. Plants from the heated greenhouse had difficulty in spiking at all. This result was interpreted to mean that with a sufficiently early sowing low temperature was not necessary, and that the crucial parameter was the time of sowing. Maksimov concluded that the day length was probably the decisive factor (Maksimov and Pojarkova 1925, 729).

7. For the tragic dilemmas of Sergei Vavilov's life see Joravsky (1965) and Kozhevnikov (1996).

8. For a more details about Vavilov's impact in Western scientific literature, see Roll-Hansen (1987). A recent assessment by a specialist in the field has been given by Harris (1990).

9. One author reports that the number of scientific institutions doing plant breeding rose from 26 in 1920 to 185 in 1932 (Revenkova 1962, 214). According to Joravsky (1970, 77) a peak was reached for the research system under the VASKhNIL in 1932–1933. It then included thirteen hundred institutions, employing altogether something like twenty-six thousand specialists.

10. See, for instance, "Ocherki istorii vozniknoveniia i razvitiia Saratovskoi oblastnoi s.-kh. op. stantsii i obshchie cherty eio organizatsii," *Trudy saratovskoi oblastnoi sel'sko-khoziaistvennoi opytnoi stantsii* (Saratov, 1923).

11. The New Aleksander Institute of Agriculture was founded near Wasaw in 1816 and moved to Pulavy, also in Poland. At the beginning of World War I it was evacuated to Kharkov.

12. For biographical information on Meister, see Guliaev et al. (1973) and Tiumiakov (1967).

13. Letter from G. K. Meister to H. J. Muller, December 1, 1934, Muller Archives, Lilly Library, University of Indiana, Bloomington.

14. For biographical and other information on Tulaikov, I have used *N. M. Tulaikov, Izbrannye Proizvedeniia, kritika travopol'noi sistemy zemledeliia* (Moscow: Izdatel'stvo sel'skokhoziaistvennoi literatury, zhurnalov i plakatov, 1963).

15. For a vivid description of the hut lab movement and other amateur agricultural science see Joravsky (1970, 54ff).

16. For a fuller description of Michurin's career, see Joravsky (1963, 1970).

17. In Russian: "toi planomernosti istoricheskogo razvitiia rastitel'novo organizma, katoraia nabliudaetsa v deistvitel'nosti."

18. Boris Aleksandrovich Keller, *Materialy k biobibliografii uchionykh SSSR, Akademiia Nauk SSSR* (Moscow and Leningrad, 1946).

19. See, for instance, Joravsky (1970, 362, fn.13).

Chapter 3.

The "Barefoot Professor"

Trofim D. Lysenko (1898–1976) was the son of a peasant in the Ukrainian province of Poltava. He did not learn to read and write until he was thirteen, and his insufficient basic education remained a serious drawback. For instance, he never learned any foreign language and was thus cut off from foreign scientific literature. Lack of formal education was only partly compensated by private study and on-the-job training. But Lysenko was intelligent and ambitious, and the Bolshevik revolution provided extraordinary career possibilities for a man of his background.

LYSENKO'S START AS A PLANT BREEDER

After attending a lower school of gardening, Lysenko was admitted to the prestigious Umansk secondary school of gardening in 1918. But the civil war prevented systematic and continuous study. In March 1922 he was appointed a senior specialist at the Sugar Trust's Belaia Tserkov selection station, located not far from Kharkov. Here Lysenko combined practical breeding work with extramural studies at the Kiev Agricultural Institute, graduating with a degree in agronomy in 1925.[1]

Two short papers by Lysenko were published in 1923 in a report from a Ukrainian conference on plant breeding. One, on tomato breeding, was authored by Lysenko alone. The other, on grafting of sugar beet, was written with a colleague. Tomatoes were to become a favorite plant in his later controversial genetic experiments and grafting one of his favorite techniques. Vegetative hybridization through grafting was a tempting speculative idea. Throughout the twentieth century, and especially during its first half, this idea was tested numerous times without any convincing positive results. But it still would not die. Today one might argue that modern knowledge about the transfer of DNA even between individuals of completely different species has confirmed the possibility of this phenomenon.

Lysenko's 1923 paper "Technique and Method for the Selection of Tomato at the Belotserkov Selection Station" is simply a detailed description of the procedure used and the aims that were pursued. The station's division of orchard (*ogorodnye*) plants had chosen cabbage, potato, and tomato as its main plants and given special attention to the breeding of tomato. Individual (pedigree) selection within existing varieties was the standard method. Hybridization was also used to combine valuable properties of different strains. Besides early ripening and resistance to pests, frost hardiness was a desired property. If new hardy strains were obtained, they would be crossed with existing strains with high-quality fruit. The description is clear and concise, perhaps somewhat excessive in detail, but quite up to the standards of good practice at the time.

The second paper, "Grafting of Sugar Beet" (Lysenko and Okonenko 1923), describes a new grafting method to speed up seed production. The authors start by explaining that with the ongoing change from mass to individual selection it has become crucially important to obtain as much seed as possible from the originally selected plant. Since the plant is cross-breeding and heterozygous, the sugar content decreases for each generation without further selection. The new method consisted in cutting up the selected sugar beet into pieces with one bud on each and grafting these pieces onto other beets. The amount of seed produced from a single beet root could thus be increased approximately tenfold. According to the paper this meant that one generation less was needed to produce seed on a commercial scale. Lysenko and his colleague argued that this new method of grafting buds was more efficient than the old one of grafting shoots.

The second paper also tells about an experiment on graft hybridization, that is, whether the hereditary properties of the scion are affected by the stock that it is grafted on. The colleague, who is responsible for the second part of the paper, does not believe in a positive result of this experiment. Lysenko, in his first part, is more open.

A method of seed treatment was also discussed, reminiscent of the vernalization that was later to make Lysenko famous. The seed of sugar beet normally need a period of rest before germination, "during which time various processes occur which lead to loss of water." The two investigators took seed a few days after harvesting, subjected them to 60 degrees centigrade for twenty minutes to speed up the loss of water, and then sowed them. They did not, however, claim much originality for this method: "Such substitution of natural rest with artificial is well known in plant physiology" (Lysenko and Okonenko 1923, 80).

These two papers of the twenty-five-year-old Lysenko are by no means spectacular in their results. Neither do they stand out as demonstrations of incompetence. They are simply quite ordinary papers within their field. The short account of the discussion that followed the presentation of the paper on sugar beet grafting indicates that Lysenko was taken seriously by other scientists in the area. Considering his peasant background and his delayed and patchy formal education, these papers represented no small personal achievement. A survey of scientific institutions for plant breeding in the whole Soviet Union lists Lysenko as responsible for garden and orchard (*ogorodnye*) plants at the Belaia Tserkov station. He appears as one of two specialists at the station. The other one was responsible for grasses (*zlaki*) (Talanov 1924, 367). Thus by the early 1920s Lysenko was already a practicing plant breeder in a senior staff position. He had taken the first steps toward a successful scientific career.

THE "BAREFOOT PROFESSOR"

Throughout his scientific work, Lysenko constantly emphasized practical usefulness. This suited both his own social and educational background and contemporary ideological fashions. But we shall also see that from very early on he had a predilection for theoretical generalization. His dream was to discover a new fundamental principle that in one stroke would revolutionize plant growing. His ambition demanded more than small contributions to existing methods and a correspondingly modest career. Lysenko, like Vavilov, belonged to what was affectionately called the "romantic" type of scientist by the Nobel Prize–winning German physical chemist Wilhelm Ostwald (1909). Lysenko had the quick mind, the enthusiasm, intensity, and dedication to his work that could inspire his coworkers and catch the fancy of the general public. He also caught the approval of Vavilov as a man who was dedicated to his science and did not waste his time in debates over Marxism and politics. Like Vavilov himself, Lysenko never became a member of the Communist Party.

After graduating from the Kiev Agricultural Institute, Lysenko in 1925 went to work at the experimental station in Gandzha in Azerbaijan, a southern country with mild winters.[2] In a *Pravda* article of August 1927, "The Fields in Winter," Lysenko was presented to the general public as a "barefoot professor." This new breed of scientist had shown how fields that used to lie fallow during the winter could be used to grow peas and oats for animal fodder. This meant not only an extra crop but also fertilization of the earth through the nitrogen fixation of the pea plants.

The article vividly depicted how Lysenko's approach to science differed from that of traditional academics. His unspoiled and sharp practical mind quickly perceived a problem, struggled with it, and intuitively grasped the solution. According to the journalist, Lysenko had started by asking: If weeds can grow on the fallow, why not just as well useful plants? He had pondered this problem whenever he was not occupied with his assigned work, read books, and used his eyes wherever he went. He had sought a principle of solution and finally found it: "Every plant needs a certain amount of heat. If everything can be measured in calories, then the task of utilizing the fields in winter can be solved on a piece of paper," that is, by simple calculation. Lysenko then started his investigations, measuring the amount of heat needed in the development of various plants. Many possibilities were tried. Another year passed with new measurements. Finally, it turned out that peas thrived in the Transcaucasian winter. As a result, "a completely new kind of life" was now opened to the local population, according to the journalist.

This romantic and unrealistic description of Lysenko's research nevertheless reflects his view of science and the impression he made on the journalist. The friendly, kidding tone indicates personal sympathy as well as harmony in the view of science. Lysenko was not a man occupied with trivial theoretical problems remote from real life. He was different from the traditional professor.

The article started by telling about another professor whom the journalist had known. His life's work was a treatise of eight hundred pages called *The Role of the Northern Bee in the Fertilization of Takveri Grapes*. The professor had died with a light conscience, confident of having fulfilled his duty. But to this very day, the journalist ironically commented, it has not been possible to make the northern bee thrive in those regions where cultivated grapes grow to ripeness. Lysenko, on the other hand, was a doer who knew instinctively to distinguish important from unimportant tasks. He was also a plain man of the people who enjoyed simple pleasures. Only once did this "barefoot scientist" smile, at the thought of his childhood cherry pie with sugar and sour cream.[3]

Perhaps Lysenko did not play his role quite as well as the journalist described it. But at least we are presented with the public expectations that met Lysenko, which he soon learned to live up to.

Lysenko's own later reminiscences confirm that it was during attempts to find varieties of legumes suitable for winter growing that he "discovered" vernalization. He noticed that some varieties of peas, which would flower quickly and ripen early in the Ukraine, developed much more slowly when similarly sown in late autumn in Azerbaijan. Other varieties that were late in the Ukraine were early in Azerbaijan (Lysenko 1936c, 19–20). From these observations of differences in the response to winter conditions he went on to experiments with wheat and other plants.

VAVILOV TAKES AN INTEREST

By 1927 Nikolai Vavilov and his colleagues at the All-Union Institute of Plant Industry (VIR) in Leningrad had become well aware of the young and energetic Lysenko. Presumably the *Pravda* article had stimulated their curiosity. In the autumn of 1927 Vavilov sent a young specialist on legumes to Gandzha to take a closer look.[4]

Vavilov's envoy was well received by an enthusiastic Lysenko, who persuaded the guest to use his bed while he himself slept on the floor. Lysenko talked continuously about his theories and results. Returning to VIR, the envoy reported that he had met an "experimenter who was fearless and undoubtedly talented, but he was also an uneducated and extremely egoistical person, deeming himself to be a new Messiah of biological science." At a following staff meeting Vavilov supported a proposal to invite Lysenko to come to work at VIR. He argued that all the best talent should be concentrated there, especially such people as Lysenko who could easily stray from the right path if left without rigorous scientific criticism. There was general agreement with Vavilov until Nikolai Maksimov, head of plant physiology at VIR, unexpectedly objected. Lysenko had poor knowledge of the scientific literature, his work was messy, and he had little interest in learning proper methods, said Maksimov. What other leading scientific institute will make a person with hardly any scientific publications into the head of a laboratory? Two more meetings were called to discuss the matter. But Maksimov continued his opposition and Lysenko received no invitation for a post at VIR (Reznik 1968, 265–67).[5]

TEMPERATURE EFFECTS ON PLANT DEVELOPMENT

Results on the aftereffect of low temperatures are presented as a by-product of a broader research program in Lysenko's first, and last, larger research publication. The aim of his *Effects of the Thermal Factor on the Duration of Phases in the Development of Plants* (1928) was to contribute to a general theory about the temperature dependence of development. In the preface he announced a second volume on "the remaining part of the experiments," which was never published.

The central idea of Lysenko's 1928 thesis was that the duration of plant development has an inverse relationship to temperature. There is a certain "sum of heat" that the plant needs to pass through a particular stage (or phase) of its development. This was a widespread and accepted idea in phenological and ecophysiological studies of the time. The "heat energy"[6] was very often measured as the "sum of heat" ("thermal time" is a modern term) in terms of "degree-days."[7] This quantity was found by summing up over a certain number of days the average temperature of each day and night. The presentation of large amounts of relatively raw data was also characteristic of the phenological literature. Lysenko's way of filling pages with masses of data was not uncommon.[8]

Lysenko made no claim of originality in searching for a law of the dependence of development on temperature. He related himself to a well-established research tradition. But so far the efforts of "physiologists, agronomists, and phenologists" had failed to solve the problem. The regularities of the annual cycles that phenologists observed indicated the existence of a simple fundamental law, and Lysenko's ambition was to find this law and develop a precise, comprehensive, and powerful theory (Lysenko 1928, 5).

In method as well as theory, Lysenko built on the work of Gavril S. Zaitsev (1887–1929). He was a friend of Nikolai Vavilov and the leader of cotton research within the VIR system.[9] Lysenko explained in the introductory discussion that he sought a generalization of Zaitsev's analysis of the development of cotton. The regularities that Zaitsev had discovered were valid only for conditions similar to those of Tashkent, where the experiments had been carried out (Lysenko 1928, 8–11).

In 1927 Zaitsev had published a paper called "The Effect of Temperature on the Development of the Cotton Plant." His method for varying the temperature was to sow at different times throughout the spring months, and thus exploit the natural variations in temperature. Lysenko used the same method in his 1928 paper. This method makes it difficult to separate the effects of light and temperature, since there is a parallel change in temperature and light regime through the annual cycle. Neither Zaitsev nor Lysenko used the experimental approach of Gassner's 1918 paper, germinating the seed at fixed temperatures and planting them in the field at different dates ranging from winter to summer. Probably neither of them had access to the necessary technical equipment for refrigeration.

Zaitsev's aim was to establish as accurately as possible the relationship between temperature and the time plants take to develop through their various phases from germination to the ripening of the fruit. As a vignette for his paper he used a quotation from Louis Agassiz: "Facts are stupid things until brought into connection with some general law." Like Lysenko, he was speculating about some fundamental law in the form of a sequence of relatively independent phases (*isofasy*) each demanding a certain amount of heat in terms of "degree-days." Zaitsev presented an "ideal curve" for the dependence of development on temperature, accompanied by a formula that described the curve, $n = 26{,}25 - 1{,}685t + 0{,}0303t^2$. Here n stands for the number of days needed to pass through a certain phase and t for the average temperature in degrees Celsius during this time (Zaitsev 1927, 31).[10]

According to Zaitsev, this formula could become "very important" for solving problems of cotton growing. For instance, evaluation of the potential of growing in various regions could be made in a more sophisticated way. Earlier it had simply been made on the basis of average summer temperatures. Now the seasonal variations could more easily be taken into account (Zaitsev 1927, 75). The possibility of predicting plant behavior, and ultimately yields, had thus been substantially improved.

Lysenko's 1928 book stressed two basic principles of high relevance to the general study of plant development. First, that development should be distinguished from growth (Lysenko 1928, 15). Lysenko is still considered a pioneer in emphasizing the importance of this distinction in the theory of plant development.[11] Second, that the duration of phases is mainly determined by temperature (Lysenko 1928, 13). This second principle becomes more and more problematic as one generalizes for more and more diverse geographical locations and climates. Among other things, it neglects variation in day length.

Lysenko pointed out that the choice of 0° Celsius as the baseline for calculations of the "sum of heat" was arbitrary. While some plants could grow and set flowers and fruit at temperatures below 0°C, others would die even at 10 to 15°C (Lysenko 1928, 5). It would be better to make the calculation from a base that represented the minimum temperature at which a certain development could proceed at all (Lysenko 1928, 16). He hoped that if the formulas and calculations were corrected in this way, the quantity of "degree-days" needed by a certain plant species to pass through a certain phase would be truly constant and the predictions would become more accurate and reliable.

Investigations of the dependence of plant development on temperature using measures of degree-days received much attention in the Soviet Union at this time. The idea of using some point other than 0°C as a baseline was also part of the common discussion. But the methods used for calculating correlations from the experimental data were sometimes problematic. Lysenko, among others, was criticized for using deficient statistical methods (Shatskii 1930).

In his 1928 preface Lysenko acknowledged that the former director of the Gandzha station had helped him with the mathematics. N. F. Derevitskii was one of the foremost national experts on the use of statistical methods in agricultural experimentation.[12] For instance, he contributed the article on statistical methods to the treatise on *Theoretical Foundations of Plant Selection* edited by Vavilov in 1935 (Derevitskii 1935). He was also an old school friend of Zaitsev who thanked him for help with the formulas for cotton development (Zaitsev 1927, 31). Derevitskii in his 1935 paper commented critically on Zaitsev's use of statistical correlations and his interpretation of the formula for the isophases (559–60). He also found Maksimov's use of statistical correlations problematic while he praised Lysenko for disproving Maximov's claim that growth and development are antagonistic (561).

Both Zaitsev and Lysenko speculated on finding simple laws underlying the regularities of plant development. But Lysenko's tendency to generalize mathematical formulas and interpret them as expressions of real natural relationships made a difference. It is reported that when Zaitsev heard Lysenko lecture on the "Effect of the Thermal Factor" in Leningrad in 1927–1928, his advice was to cut the mathematics in order not to obscure the valuable phenological observations.[13]

With the formula $n = A/(t - B)$,[14] Lysenko summed up what we can call his early theory of stages in the development of plants. This fomula states that the duration of each phase is inversely proportional to the amount by which the temperature exceeds a certain minimum (B). And A is a constant for each plant and phase representing the "sum of heat" required for passing through that specific phase (Lysenko 1928, 17).

In fact, this formula suits only the cases and intervals where increasing temperature speeds up the process. As already mentioned, the need for low temperature (i.e., temperature *below* a certain maximum) for certain stages in certain grain varieties was also briefly described in Lysenko's 1928 book. Such a requirement for reduced temperature could not directly be subsumed under the given formula. But Lysenko sketched a program for further research to accommodate the cold requirement under the general theory. The common principle of this research program was that the lawlike dependence of development on temperature "is qualitatively the same for all plants, only quantitatively is it different" (Lysenko 1928, 130). He called this his "a priori" assumption (Lysenko 1928, 8).

The requirement for low temperature can be formally accommodated by Lysenko's formula if one introduces the "sum of cold" as equivalent to a negative "sum of heat." If $A' = -A$ is a cold requirement, the formula becomes $n = A'/B - t$. The amount of "cold," accumulated on days with temperatures *lower* than a certain given limit, is the decisive quantity. In late 1928 Lysenko presented this formula for a cold requirement at a scientific meeting in Kiev organized by the Sugar Trust (Lysenko 1929).

In evaluating Lysenko's *Effects of the Thermal Factor* one must consider the state of the research tradition in environmental physiology, including phenology, to which he belonged. The physical and biological interpretation of constants such as Lysenko's A and B was problematic, but not necessarily meaningless or unscientific. A meteorologist who criticized Lysenko for weak statistics overshot the mark when he declared that "so far the temperature sum, as a climatic index, has no objective foundation," demanding that agricultural meteorology should be left to the agricultural meteorologists. After all, he admitted himself that there is an "incontestable connection" between the rate of development and the temperature conditions, and even that this connection can be "determined with accuracy" in laboratory experiments (Shatskii 1930, 366ff).

The results of Lysenko's investigations of the relationship between temperature and development were by no means spectacular, but they should not be underestimated, either. His ability to predict the behavior of plants under new climatic conditions, and thus to help in choosing the right variety for a certain region (Lysenko 1928, 124), was most likely exaggerated. But as part of the ongoing activity within an active field of research, Lysenko's work upheld acceptable standards. He was a respected member of the research community within his chosen field.

Both the methodological and the theoretical sides of Lysenko's book on temperature effects have weaknesses that reveal his lack of formal training. His methodological immaturity is clearly seen in an uncritical attitude toward his own results and arguments. For instance, he writes that the "coincidence of theoretical data with factual observations is almost complete" (Lysenko 1928, 19), and that "[o]ne may say that the theoretically calculated data are more correct for the particular year and the particular region than those which are obtained in practice. They speak not of one hectare with six repetitions but of the whole region with its definite climatic conditions" (Lysenko 1928, 124–26). Such statements indicate a conflation of theory with observation and speculation with experience. Lysenko also showed a tendency to neglect alternative explanations of the observed data. To be fair, he admitted that other factors, such as the length of the day and the intensity of light, could affect development. But he insisted that temperature was the "main factor" (127). As we shall see in the following chapters, strong confidence in the truth of theories that had been only superficially tested was typical of Lysenko's scientific behavior.

Lysenko did not work only with questions of plant physiology during his Gandzha period. He also pursued problems of selection and inheritance. The concluding discussion of his 1928 book sketches a broad perspective in which developmental effects of temperature are linked to plant breeding. The making of early-ripening varieties had high priority in Soviet plant breeding at the time. This was considered an important means of pushing agriculture to the north as well as avoiding the effects of summer drought in southern parts. Lysenko argued that the determination of thermal constants of development for different varieties could help selection for early ripening. He suggested that these constants represented fundamental hereditary properties that were independent of particular climatic conditions (Lysenko 1928, 129).

LENINGRAD CONGRESS, JANUARY 1929

A grand event in Soviet agricultural science was the All-Union Congress on Genetics, Plant Breeding, Seed Production, and the Raising of Pedigreed Livestock in Leningrad, January 10–16, 1929. Vavilov was the main organizer of this meeting attended by more than a thousand participants from all parts of the Soviet Union plus some prominent foreign plant breeders and geneticists, including Erwin Baur and Richard Goldschmidt from Germany and Harry Federley from Finland. The congress was also a public demonstration of science as a main force in building the new socialist society. The top party leader in Leningrad, Sergei Kirov, welcomed the participants at the opening session.

The main lecture on the physiology of plant development, "Physiological Control of the Length of the Vegetative Period," was given by Maximov. He began by mentioning the big practical challenges of Soviet agriculture and the constant complaints about the lack of practical results from research in plant physiology. An acute practical problem of Soviet agriculture was the destruction of winter cereals by cold winters with little snow. During the winter of 1927–28 as much as 90 percent of the plants had been killed in some regions of the Ukraine. It was the worst situation in thirty-five to forty years. A special conference was held in Kharkov in July 1928 where plant scientists like Vavilov, Meister, and Liubimenko discussed possible countermeasures.[15]

Maksimov assured the audience that the lack of practical results was not due to a lack of goodwill on the part of the scientists. And he ended with the optimistic claim that an elucidation of the factors that determine the length of the various stages of plants would also make it possible to direct their development at will. "With the mastery of these causes we will have in our hands the possibility of calling forth fructification in plants at any time we wish." The practical economic importance of such scientific achievements would be enormous and "needed no further elaboration" (Maksimov 1929a, 19).

A newspaper report on the following day, the last day of the congress, emphasized the practical implications of Maksimov's lecure. "It is possible to convert winter cereals into spring cereals," the readers were told. This feat had already been carried out under laboratory conditions. However, it was not practical to transfer these specific methods to field conditions, said Maksimov.[16]

The content of Maksimov's lecture was purely theoretical. Besides photoperiodism, he discussed in detail the "aftereffect of temperature," including Lysenko's formula for cold requirement. Lysenko's attempt to formulate "a quantitative law" was undoubtedly of great interest, said Maksimov, but it was still based on too few experiments (Maksimov 1929b, 175).[17] Maksimov also argued, against both Gassner and Lysenko, that day length rather than a period of low temperature was the most important factor in determining the timing of development (Maksimov 1929a, 6–8). Furthermore, he stressed the general lack of precise experimental data about factors influencing development. The inadequate technical equipment of his laboratory was a block to such research. With the existing data, all attempts to express the relationships in strict mathematical formulas were of doubtful value, concluded Maksimov (1929a, 19), clearly addressing Lysenko.

Lysenko presented his paper, "On the Nature of Winter Annual Plants," later in the same day, and Maksimov participated in the discussion.[18] On his registration form Lysenko had written that he wanted to participate in the section for selection, but his paper was placed in the section for study of cultivated plants together with other studies of temperature effects.[19] It has been reported that Maksimov wanted to reject the paper when it was discussed in the organizing committee. But Vavilov had insisted that though the paper was weak Lysenko's work was original and promising. Vavilov also argued that full participation with a paper would be a useful experience for the young scientist (Reznik 1968, 268).

Lysenko's paper was theoretical as well as experimental. He argued that low temperature was a necessary condition for flowering and criticized Maksimov's view of light as the crucial factor. In his experiments Lysenko had buried slightly germinated or merely swelled seed of winter wheat in snow and sown them in the field at regular intervals from early March until late April. He found that swelled seed of the winter wheat variety Kooperatorka, which had been chilled in the snow for thirty-eight days or more, spiked normally. Plants from untreated seed chilled for the same periods and sown at the same time did not spike.[20] Lysenko thus continued to develop the quantitative aspects of his investigations of the cold requirement.

The decline and subsequent rise in frequency of spiking, according to date of sowing, was taken by Lysenko as a proof of his view that temperature was more important than day length. For the first sowings the natural low temperatures in the field provided sufficient cold conditioning. Later sowings were dependent on artificial cold treatment for flowering (Dolgushin and Lysenko 1929, 194). An alternative interpretation would be that flowering was promoted by the short days of early spring.

Lysenko concluded his paper by arguing that the case fit his general theory of stages in the development of plants. For instance, his principle of a fundamental separation of growth from development was confirmed: the essential processes in the cold-requiring first phase could not be linked to any visible morphological changes, and development continued even if growth was stopped by low temperature. Indeed development in this phase was more rapid the lower the temperature, as opposed to growth processes. The second phase, spike formation, confirmed the principle of separate and independent phases. In contrast to the first, this second phase proceeded more quickly the higher the temperature was (Dolgushin and Lysenko, 1929, 198–99).

The possibility of practical application was not discussed in Lysenko's paper. This was, however, discussed by another speaker, a professor at the agricultural institute in Kiev, the institution where Lysenko had taken his degree a few years earlier. Unfortunately, Maksimov's laboratory had investigated effects only on young plants and not on seed, observed this professor. To transfer small plants into the field poses great technical problems. With merely swelled seed an ordinary sowing machine can be used (Tolmachev 1929, 539). The main theme of this paper given by the Kiev professor was a biochemical theory of flowering, similar to earlier ideas of Maksimov.[21] Flowering was seen as dependent on the balance between synthesis and breakdown of organic substances in the plant. In spring varieties the "stem plasm" was ready for heading, while in the winter varieties a further breakdown process was needed to reach the right balance. This could be achieved in weakly germinated seed at low temperatures (Tolmachev 1929, 542–43).

Maksimov's laboratory had for some time been working on the aftereffect of low temperature and its possible practical application. Together with a collaborator, Maksimov soon published another paper with an extensive discussion of both Tolmachev's and Lysenko's results. A long footnote pointed out that it was well known in nineteenth-century Russian agriculture that cold treatment of slightly germinated seed made it possible to sow winter wheat in the spring. But here as in other countries this knowledge had later been forgotten (Maksimov and Krotkina 1929–30, 432–33). Maksimov saw Lysenko's experiments mainly as a "confirmation and further development" of earlier results obtained by Gassner, Maksimov, and others. It was primarily the theoretical interpretation that differed.

In Maksimov's view Lysenko's division of development into separate phases was "formal" and lacked explanatory value. The fundamental difference between winter and spring wheat is not higher or lower temperature constants, that is, larger or smaller need for low temperature, but the presence or absence of a "braking" factor, argued Maksimov. If the factor was lacking, one had a pure spring variety. Maksimov was thinking in terms of a mechanism of biochemical balance rather than an active hormone (Maksimov and Krotkina 1929–30, 469–70). Lysenko on his part was skeptical of reductionist chemical explanations whether in terms of hormones or biochemical balance.

Altogether Lysenko's appearance at the Leningrad congress in January 1929 and the responses that he got confirmed his rising scientific status. He received positive attention from the leading experts within the specialty; a dose of criticism that is included is quite normal in science. Lysenko was now in a position to make a respectable career in Soviet agricultural research. A less ambitious person would have good reason to be well satisfied. His situation was quite different from that of the dreamers and amateurs in the "hut lab" movement. Lysenko had studied at a respectable agricultural institute. He was employed in a governmental research station, and he was in dialogue with the leading experts in his field.

Like other highly ambitious young scientists, Lysenko had apparently hoped that his Leningrad paper would create a big stir in the learned world. When the result was only moderate attention, he was disappointed. Later hagiographic accounts of Lysenko's life (Popovskii 1948; Dolgushin 1949) vividly describe Lysenko's righteous indignation at the condescending and arrogant behavior of leading academic scientists. These accounts are obviously marked by hindsight and a later need for legitimation. Nevertheless, they indicate that the Leningrad congress of January 1929 marked a watershed in Lysenko's career. From now on he was set on a collision course with the scientific establishment and traditional science.

FROM WISHFULNESS TO CYNICISM

The wishful character of the young Lysenko's scientific reasoning derives from his lack of scientific education as well as his personality. At the start he was a gifted but poorly educated enthusiast. The heroic picture drawn both by his disciples and by official biographers ties in well with that of other sources. The Australian plant physiologist and diplomat Eric Ashby, for example, in his 1946 book, *Scientist in Russia*, makes Lysenko less heroic, but hardly less sympathetic:

> He is not a charlatan. He is not a showman. He is not personally ambitious. He is extremely nervous and conveys the impression of being unhappy, unsure of himself, shy, and forced into the rôle of leader by the fire within him. He believes passionately in his own theories. . . . (Ashby 1946, 116)

The Russian geneticist A. A. Sozinov has confirmed this picture of the young Lysenko as a sincere enthusiast. He was a friendly and modest man who sincerely believed in his theories. Only when someone criticized his theories did he become aggressive. However, Lysenko's poor education made it impossible to discuss with him, according to Sozinov. He would always find some arbitrary argument that he claimed to be relevant.[22]

The young Lysenko appears as modest and aloof, socially uncomfortable among the well-educated middle-class academics who dominated the scientific community. He was also an enthusiast who did not spare himself. His intense engagement in work and apparent oblivion of his own personal gains or well-being, together with a modest and socially insecure demeanor, evoked sympathy among his scientific opponents. Eleanor Manevich, who was a genetics student in the late thirties, told half a century later about a brief encounter with Lysenko in the cloakroom after a meeting at the 1939 conference (or perhaps it was in 1936?):

> We were standing there. He did not look well and was very hoarse. I felt pity for him. He had been criticized by all the geneticists. I

said to him: "Trofim Denisovich you should take care of your health." His overcoat was not very warm.[23]

The picture of the young and well-meaning enthusiast is not incompatible with Lysenko's later cunning and ruthless use of power against scientific opponents. Messianic enthusiasm for one's own ideas has often inspired a suppression of opponents in the name of truth.

Lysenko was also an intelligent man who soon learned to see through the Stalinist system. He grew cynical and calculating with increasing age, experience, and power.[24] The influence of his ideological aide Isaak I. Prezent no doubt contributed to such a development. And the ability of senior scientists such as Vavilov and Sapegin to restrain Lysenko's wild speculations dwindled as his prestige and status grew through the 1930s.

NOTES

1. For details on Lysenko's early life see Soyfer (1994, 8–12).

2. The experimental station in Gandzha was established around 1925. Breeding of cotton for irrigated areas of Azerbaijan and the development of other local cultures were main tasks of this station (Pisarev 1929, 223).

3. *Pravda*, August 7, 1927, p. 6.

4. The following account of Vavilov's attempt to give Lysenko a laboratory at VIR is based on S. Reznik's biography of Vavilov (Reznik 1968). Reznik interviewed a number of the participants who were still alive. Among them was Vavilov's special envoy, Nikolai R. Ivanov (1902–1978), who later became a professor of legumes. He is the main source of the account that Reznik (1968, 262–67) gives of this contact between Lysenko and VIR.

5. The timing and character of the first contact between Lysenko and Vavilov has been the subject of some controversy. Without being very specific, Mark Popovskii, Russian journalist and science writer, has claimed that Vavilov came into contact with Lysenko in 1927–1928 and that he helped Lysenko with experimental materials and equipment. Popovskii has suggested that Vavilov first met Lysenko on a visit to Gandzha during this period (Popovskii 1966, 13–14). But the account of Popovskii has been

sharply criticized by Zhores Medvedev for giving the impression that Vavilov was a main promoter of Lysenko's early career. It is quite unlikely that Vavilov visited Gandzha during this period, according to Medvedev, because he was busy traveling elsewhere and there is no independent evidence of such a meeting. Furthermore, Popovskii's account of how the unknown Lysenko was discovered by Vavilov is absurd simply because Lysenko was by 1926–1927 already a well-known person in agricultural science (Medvedev 1967, 228). The Russian historian of science V. Esakov has claimed that Vavilov met Lysenko at Belaia Tserkov before he moved to Gandzha in 1925 (personal communication, Moscow, September 21, 1996). Popovskii later dropped the suggestion that Vavilov first met Lysenko at Gandzha but upheld his picture of Lysenko as a protegé of Vavilov (Popovskii 1983). Reznik's (1968) account of Vavilov's attempt to move Lysenko to VIR does not contradict Popovskii's general interpretation of their early relationship. According to Dubinin (1990, 53) Vavilov first met Lysenko on a visit to Gandzha in 1928.

6. Lysenko uses the term *teplovaia energiia* (1928, 5).

7. More sophisticated versions of the same kind of measure are still used in plant physiology describing and analyzing the development of phases.

8. See, for instance, a phenological study on wheat carried out in 1926–1927 in Oregon and published in a quite prestigious context as late as 1948 (Nuttonson 1948).

9. After graduating from the Petrovka in 1914 Zaitsev had moved to central Asia to become one of the world's leading specialists on cotton. A close friendship with Vavilov, started in 1916, is described in Reznik's biography of Zaitsev (1981). A number of Zaitsev's most important papers have been republished (Zaitsev 1963, 1980).

10. Zaitsev's formula appears not quite to correspond to the curve that he draws. But at least it indicates a similar curvilinear form.

11. Conversation with Ola Heide, professor of plant physiology at the Agricultural University of Norway, September 15, 1991.

12. Nikolai F. Derevitskii (1882–1959) worked at experimental agricultural stations in the Ukraine from 1918 to 1925, when he became director at Gandzha. From 1927 to 1932 he worked in Tashkent, from 1949 in Moldavia. See N. F. Derevitskii, *Opytnoe Delo v Rastenievodstve* (Experiments in plant cultivation) (Kishinev: Academy of Sciences of the Moldavian SSR, 1962).

13. Conversation with Zaitsev's daughter Maria G. Zaitseva in Moscow,

April 27, 1990, at the Timiriazev Institute of Plant Physiology. Her sources were mainly Zaitsev's coworkers. Zaitsev died suddenly from appendicitis in Moscow in January 1929, on his way to the big conference on genetics and breeding in Leningrad.

14. Lysenko proposed the basic formula $A + Bn = St$ for calculating the duration of the phases. Here B represents the minimum temperature for a certain phase to proceed. And A is the "sum of heat" (at temperatures above B) needed for passing through this phase. The mean temperature (of one night and day) is expressed by t, and St stands for the sum of mean daily temperatures over the whole period in question. Lysenko emphasized that this is, of course, not to be understood as an energy equation, but merely as a relationship between temperature and time measurements. According to this formula, the number of days needed for passing through a certain phase is $n = (St - A)/B$. If we now express St as tn (mean temperature for the whole period multiplied by the number of days) the formula becomes $A + Bn = tn$ and the number of days $n = A/(t - B)$. (Lysenko 1928, 18.)

15. Vavilov lectured on "Botanico-Geographical Deliberations about the Possibilities of Extending the Cultivation of Winter Wheat in the SSSR," Meister on "Perspectives in the Selection of Winter Wheat," Maksimov on "Achievements and Perspectives in Investigations of the Physiology of Winter-Hardiness," and Liubimenko on "The Necessity of Ecological Investigations of Cultivated Plants." The lectures were published in *Gibel' ozimykh khlebov i meropriiatiia po eio preduprezhdeniiu* (The perishing of winter cereals and means to avoid it), *TPBGS* (Moscow-Leningrad) supplement 34 (1929).

16. "Uspekhi sovetskogo nauka" (Achievements of Soviet science), *Leningradskaia Pravda*, January 16, 1929, p. 3.

17. This comment on Lysenko's work was not included in the shortened version of the lecture published in the proceedings of the conference (Maksimov 1929a). It is found in the German version (Maksimov 1929c).

18. Minutes of the section on cultivated plants. TsGANTD–St. Petersburg, f. 318, o.1, d. 248, l.14. See also Manevich (1991, 139).

19. On the application form, Lysenko's entry for the section of plant breeding (*selektsionnuiu*) is crossed out in pencil and study of cultivated plants (*izuchenie kul'turnykh rastenii*) is written in. TsGANTD–St. Petersburg, f. 318, op. 1, d. 255, l.161.

20. Lysenko's main experiment was conducted as follows. Lots of winter barley (Pallidum 419) and winter wheat (Kooperatorka) were divided into three parts. One was "weakly germinated," the second was

only "swelled," and the third was not treated and served as control. On March 2 the treated seed were put into sacks and kept in a heap of snow. From March 4 seed were sown with intervals of a few days. Since the two kinds of treatment did not make any difference to the result, they are not distinguished in the report. Untreated seed were sown both in the laboratory and in the field as control series (Dolgushin and Lysenko 1929, 191).

For the winter wheat the result was that all treated plants sown in the first week headed normally. The controls also headed to some extent, even the controls sown in the laboratory. With later sowing the heading of treated plants was reduced while that of the untreated ceased completely. After some time the heading of plants from treated seed again increased. Those sown on April 9 and later headed normally. Lysenko's conclusion was that once germinated or swelled seed had received the necessary period of low temperature, the plants would spike whenever they were sown. This period was calculated to be thirty-eight days for the winter wheat and twenty-eight days for the barley (Dolgushin and Lysenko 1929, 197).

21. See, for instance, Maxsimov and Pojarkova (1925, 729).

22. Interview with A. A. Sozinov at the Institute of Genetics of the Academy of Sciences, Moscow, December 11, 1986. Sozinov's father was a professor at the Agricultural institute in Odessa and his mother a research worker at Lysenko's institute. He became director of the Institute of Genetics of the Academy of Sciences in the 1980s.

23. Conversation with Eleanor D. Manevich in her apartment in Moscow, December 10, 1992.

24. One of Lysenko's most persistent opponents, the geneticist V. P. Efroimson, has confirmed the picture of the mature Lysenko as a cynic rather than a fanatic. He understood well what was going on in the Soviet Union and would privately run down everything and everybody who obstructed his purposes. (Conversation with V. P. Efroimson in his apartment in Moscow, December 6, 1986).

Chapter 4.

Marxism and Science Policy

Stalin proclaimed 1929 as the year of "the great break." The economic and social revolution that had slowed down in the early 1920s was now resumed at full pace. After one step back, to the so-called New Economic Policy (NEP), the regime took two steps forward, to the forced introduction of socialism. A central system of planning was established, and two grand economic schemes were launched: the building of a modern heavy industry and the collectivization of agriculture. While the first was successful in the short run—it created the industrial basis for the defeat of Hitler—the effect of the second was mostly negative. It destroyed traditional rural society but inspired little renewal, and it had a destructive effect on agricultural production.

The period of the first five-year plan, 1928–1932, was also a period of cultural revolution. Research and higher education were abruptly expanded. Students with peasant or worker backgrounds were recruited in large numbers. There was a proliferation of crash courses to make up for poor prior education. People with the appropriate proletarian background were also systematically preferred for positions in administration, teaching, research, and so on. There was a special term for those who were promoted in this way: *vydvizhenets* (pushed up). Radical methods of teaching were tried. But by 1932 most of the educational experiments were called off. As in art and literature, there was a return to ideals that were conservative or even reactionary by liberal bourgeois standards.

Economic modernization for the purpose of increased production was a primary aim of the great break. Economic efficiency, in a Soviet-Marxist interpretation, became a superior criterion of evaluation in all fields of public policy. The superiority of the socialist economic system was not to be disputed. But within this framework, when ideological orthodoxy came into conflict with perceived economic efficiency, the latter had a chance to prevail.

Catastrophic famines had followed the civil war. Scarcity of food was a chronic headache for the new government of a country that had once been the main granary of Europe. Increasing agricultural production was a top priority. The backwardness of Russian agriculture in both technology and economic structure made modernization through a new form of collective ownership, permitting large-scale mechanized operations, an appealing prospect. Science appeared as a crucial factor in this comprehensive modernization program. Collectivization aimed at a radically new economic structure with opportunities for a systematic use of science that existed nowhere else in the world. Agriculture appeared to be particularly well suited for a thoroughly planned and centrally directed system of scientific research and development. But such utopian dreams crashed against the harsh realities of the Soviet Union. New famines with millions of deaths followed collectivization in the early 1930s. Agricultural science was caught up in the turmoil created by the suppression of the *kulaks* (prosperous peasant farmers). The most prosperous part of the peasantry was also the most advanced technologically and economically. Their elimination by execution and deportation was a heavy blow against the modernization of Russian agriculture that was well under way before the 1917 revolution. Collectivization destroyed much of the competence and infrastructure that had been built up during the preceding decades.

PLANNING OF SCIENCE

According to Nikolai Bukharin's vision of science, it was "necessary courageously and decisively to break with old academism." The days of "scholastic monasteries, the laboratories of the alchemists and the quiet offices of individual university scholars" were gone, he declared to an international audience in London in the summer of 1931 (Bukharin 1931a, 326). Science now had to be organized on the principles of big industry, as was already the case in the most advanced capitalist countries in the preceding century. But he conceded that the principles of economic planning could not be directly transferred to scientific research. The special character of scientific research demanded special forms of planning.

Nikolai Vavilov was the central scientist in this planning work. He deplored the continuing separation of agricultural research from practice. He agreed with Stalin, Bukharin, and other Marxist ideologues who stressed that science had become sterile because it lacked contact with practical work. But Vavilov also pointed in the opposite direction. He pointed to the existence of a fund of theoretical knowledge that had not been utilized (Vavilov 1931a). On Vavilov's account the cause of this deficiency lay in practical agriculture no less than in science. However, as we shall see, it was the perspective of Bukharin and Stalin that came to dominate science policy in the near future. When the promises of great increases in agricultural production, held out by Vavilov, Tulaikov, and other leading scientists, did not materialize, it appeared obvious to politicians and administrators that something was wrong with the way science was organized and conducted. The confidence in science as such was almost boundless. If it was only done in the right way, all practical problems could be solved.

At the conference on planning of science in April 1931, Vavilov enthusiastically described how the system for agricultural research had been thoroughly renewed by establishing the Lenin Academy of Agricultural Science (VASKhNIL). For instance, there now existed thirty-five "branch institutes" linked directly to the practical demands of particular branches of agriculture. Each of these institutes had a network of "zone stations" and "support points" spread throughout the Soviet Union. In addition to the "branch institutes," there was a set of central disciplinary institutes taking responsibility for the development of methods, for scientific quality control, for the guidance of the branch institutes and their subordinate system, and for long-term research (Vavilov 1931a, 137).

Vavilov fanned the public's expectations, which were already very high: "There is no doubt whatsoever that in the shortest time this new system will eliminate the discrepancy between research work and the demands of life that our agriculture faces" (Vavilov 1931a, 138). These words Vavilov had to retract a few years later as he stepped down from his post as president of VASKhNIL. But for the time being, agricultural research was presented as a general model for science. Vavilov concluded his talk by proposing the *"immediate creation of a central organ for planning and evaluation of scientific work in the whole Soviet Union"* (Vavilov 1931a, 138).

PHILOSOPHICAL IDEAS BEHIND SCIENCE PLANNING

Less than three months later, Bukharin led the Soviet delegation to the Second International Congress of the History of Science in London, June 29 to July 4, 1931. A large and high-powered delegation arrived spectacularly by special airplane at the last moment. However, the Soviets got much less time on the official program than they wanted. An extra side session was improvised. And only ten days after the end of the congress, a book appeared containing full versions of all their talks (Bukharin et al. 1931). Both the efficiency and the content of their performance greatly impressed a group of young and politically radical British natural scientists, stimulating their inclination toward Marxist views on science and society. Among them were rising stars of natural science such as John Desmond Bernal, Joseph Needham, and J. B. S. Haldane, who later, for a shorter or longer period, came to take a favorable view of Lysenkoism.

At the London congress Bukharin spoke on the unity of theory and practice in scientific research. That science should represent the interests of the working class and have "practice" as the fundamental criterion of scientific truth were core principles of his doctrine. The ideal of a new "proletarian" science derived from Marxist sociology of knowledge and the "unity of theory and practice" was inspired by a pragmatic conception of knowledge.

Before the 1917 revolution Bukharin had been a pupil of the medical doctor and Bolshevik philosopher Aleksandr Bogdanov, who is best known as Lenin's rival to political leadership of the Bolsheviks. Bogdanov was the main target of Lenin's most famous philosophical tract, *Materialism and Empiriocriticism* (1908). Bogdanov's influence derived in large measure from his responsibility for the schooling of party cadres. In the years following the 1917 revolution he inspired the *proletkult* (proletarian culture) movement until Lenin closed it down because it was becoming uncomfortably popular and also too militant for the ideological moderation of the NEP period.[1]

Bogdanov was a follower of the "empiriomonism" of the Austrian physicist and philosopher Ernst Mach. It was a type of radical empiricism aiming to surmount the contradiction between materialism and idealism by reducing all knowledge to elementary "sensations" (*Empfindungen*). The term "monism" announced a break with the traditional dualism between mind and matter and was popular among progressive and scientifically minded thinkers up through the first half of the twentieth century. Lenin demanded more respect for the existence of real objects independent of the human mind. He accused Bogdanov of dissolving the materialist basis of Marxism.

In 1909 Bogdanov published an essay called "The Philosophy of Contemporary Natural Science" in a collection called *Outline of the Philosophy of Collectivism*.[2] He claimed that twenty to thirty years earlier, progressive natural scientists had been materialists but now they had become "empiriocriticists." The reason for this change was that the individual researcher had been replaced by the research collective. Modern laboratories are nothing else than "factories of knowledge about nature," he declared (Bogdanov 1909, 37–38). Mach's epistemology had liberated science from the dualism of matter and spirit and all the accompanying contradictions.

Together with the new industrial organization of science, this epistemology provided a unified conception of science in its social setting, according to Bogdanov.

Darwinism was a main source for Mach's idea that scientific theories should be seen as instruments for the "economy of thought" rather than as referring to independently existing objects. From this view Bogdanov derived his theory of science as a product of and an instrument for socially organized work. He wanted to demystify science through a reduction of its content to the ordinary experience of workers. As an example he described how astronomy from Babylonian times to the Renaissance was slowly constructed by people who collected and systematized practical experience (*opyt*). It is clear, declared Bogdanov, that "science is nothing but *the organized experience of human society*" (Bogdanov 1918, 3). A "general theory of organization" was to give the final answers to questions concerning the "secret of science" and its role in a socialist society (Bogdanov 1918, 92).

Bogdanov's philosophy of science belonged to the neopositivism typical of the early twentieth century. With some reason Lenin pointed out its negation of materialist metaphysics. Bogdanov is important in our account because he showed how Mach's radical empiricism could be turned in a socialist direction. His view was reductionist in the sense that he saw the knowledge of natural science as built from the practical experience of ordinary workers. It could be called "social constructivism" in the terminology of the 1990s. His ideas were attractive as an epistemological underpinning in the struggle to create a truly "proletarian" science. We can recognize Bogdanov's influence on Bukharin's thinking in the early 1930s. But the views of Bogdanov were no longer politically correct, and Bukharin's writings lack references.[3]

Bukharin's ideas about the planning of science assumed a fundamental difference in the organizing principles of capitalist and socialist societies. While capitalism develops through blind and fortuitous causal processes, socialism develops according to an aim. In socialism "the future lies ahead as a plan, an aim: causal connection is realized through social teleology." One consequence of this was the possibility within a socialist society to fuse "theory and practice" and to eliminate "the rupture between intellectual and physical labor" (Bukharin 1931b, 29–30). The ideal goal of socialist planning was, in other words, a society where teleology and efficient causation coincided, that is, where everything happened in accordance with the great plan.

The more specific plan for science that Bukharin adumbrated was also derived from the vision of a socialist and rationalist utopia where purposes are in full harmony and causal connections are totally transparent. Imperfect knowledge, lacking rationality in action, and conflicting interests did not worry him much. Bukharin outlined a truly utopian program for education and scientific research:

> ... gradually destroying the division between intellectual and physical labour, extending the so-called "higher education" to the *whole* mass of workers, Socialism fuses theory and practice in *the heads of millions*. Therefore the synthesis of theory and practice signifies here a quite exceptional increase in the effectiveness of scientific work and the effectiveness of socialist economy as a whole. The unification of theory and practice, of science and labor, is *the entry of the masses* into the arena of cultural work, and the transformation of the proletariat from an object of culture into its subject, organizer, and creator. This revolution in the very foundations of cultural existence is accompanied necessarily by a revolution in the *methods* of science. . . . (Bukharin 1931b, 31–32)[4]

Bukharin's prime example of true revolutionary unity between theory and practice was Soviet agriculture and agricultural science: "One can feel with one's hands how the development of socialist agriculture pushes forward the development of genetics, biology generally, and so on" (Bukharin 1931b, 16).

Nikolai Vavilov also participated in the Soviet delegation to the London congress. Like Bukharin, his ultimate aim was to "master the historical process." Speaking about the origin of cultivated plants, Vavilov expressed a robust belief in the practical usefulness of history.

> By knowledge of the past, by studying the elements from which agriculture has developed, by collecting cultivated plants and domestic animals in the ancient centers of agriculture, we seek to master the historical process. We wish to know how to modify cultivated plants and domestic animals according to the requirements of the day. We are but slightly interested in the wheat and barley found in graves of Pharaohs of the earliest dynasties. To us, constructive questions—problems which interest the engineer—are more urgent. (Vavilov 1931b, 97)

We can see from Vavilov's words that history was not his primary interest. Vavilov expressed himself without the Marxist rhetoric of Bukharin. He just provided an example to illustrate Bukharin's view. Bukharin explained how Marxism had taught man how to master the historical process through its unification of theory and practice. Vavilov demonstrated how man was learning to control "the historical process" in a particular field, namely, by "directing the evolution of cultivated plants and domestic animals according to our will" (Vavilov 1931b, 106).

"Unity of theory and practice" was the ubiquitous catchphrase of Soviet science policy debates in the 1930s. Critically pointed against the "ivory tower"[5] academic tradition, it appealed to practically minded people tired of academic hairsplitting. It was widely accepted, also among scientists who had little sympathy for Marxist doctrine, that academic science had been too isolated from practical life, and that a change was needed. The demand for a more practical orientation of science fed the popularity of the so-called practice criterion of truth. This criterion said roughly that a theory is true if it leads to practical success. But "practical success" is an ambiguous term. It could cover such different things as confirmation by scientific experiments, technical feasibility, economic efficiency, and political victory.

To clarify the meaning of the phrase "unity of theory and practice" Bukharin quoted from the second and eleventh Feuerbach theses of Karl Marx:

> The question of whether objective truth can be attributed to human thinking is not a question of theory but is a *practical* ques-

tion. In practice man must prove the truth, that is, the reality and power, the this-sidedness of his thinking. The dispute over the reality or nonreality of thinking, which is isolated from practice is a purely *scholastic* question. (509)

Philosophers have *interpreted* the world differently; it's more important to change it.[6]

Bukharin explained: "The problem of the external world is here put as the problem of its transformation: the problem of the cognition of the external world as an integral part of the problem of transformation: the problem of theory as a practical problem" (Bukharin et al. 1931, 15–16). For Bukharin the Feuerbach theses of Marx implied a criterion of truth that makes scientific truth dependent on "practical" social aims. The goal of proletarian science was to change the world, not to understand it. Thus the practice criterion for the truth in science was not a Stalinist invention. It was part of the heritage of the Marxist philosophical tradition.[7]

The theorizing of Bukharin and the way it was echoed by Vavilov indicates that Marxist philosophy played an important role in Lysenko's rise, not by forming the content of his theory, but by affecting the general standards of argument. Science policy was a vehicle for this philosophical influence. A crude instrumental view of science and the accompanying demand for economic efficiency came to dominate Soviet science policy. A simplistic pragmatic view of scientific knowledge with origins in a radical empiricism and social constructivism in the spirit of Bogdanov was the Marxist inheritance that formed the Soviet theory of science as it materialized in science policy. The sophisticated interpretations of dialectical materialism as applied to scientific method played a subordinate role. It could certainly serve as a source of inspiration for some individual scientists, as Loren Graham (1972, 1993) has claimed. It could also be a useful means of protecting controversial science against philosophical charges from orthodox Marxists. But by focusing so exclusively on the positive scientific effects of an advanced version of dialectical materialism, Graham neglects the main influence of Marxist philosophy on Soviet science—through the principles for science policy as expressed by Bukharin and echoed by Vavilov.

BIOLOGY AND PHILOSOPHY

In the Soviet discussions over natural science and philosophy in the 1920s, biology was a key area, and Lamarckism versus Mendelism became a key issue. The emphasis of Lamarckism on the interaction of organism and environment, including an immediate and directed influence of environment on heredity, made it attractive to a general dialectical view of nature that was preoccupied with the interdependence of the whole and its parts. Mendelian genetics, with its concept of stable genes that moved unchanged through time and different environments, appeared less germane. However, the logical links were not strict. Mendelian genetics could also be well accommodated within the framework of dialectical materialism. When philosophical debate hardened and the scope for pluralism narrowed at the end of the 1920s, classical Mendelian genetics at first prevailed over Lamarckism.

The main conflict ran between a "dialectical" and a "mechanistic" view of natural science. A group of young geneticists headed by Aleksandr S. Serebrovskii (1892–1948) fought on the side of dialectics under the leadership of the philosopher A. M. Deborin. The outcome was a victory for dialectics and for Mendelism. It could also be called a victory for science over abstract philosophy, because the young geneticists were able to integrate the conception of the stable gene as a basic premise in a dialectical view of biology.[8] We now know that Lamarckism was a theory with little future. It would have been a setback for science if classical genetics had been suppressed at this point. But for science in general and genetics in the longer run, this philosophical victory was no blessing. It gave precedence for the authority of philosophy to intervene into scientific questions.[9]

The "mechanicists" defended a scientific view that emphasized the autonomy and authority of science in relation to philosophy.

They were materialistic reductionists in the sense that all phenomena in the world, including the biological and the social, must in the last instance be explained through physicochemical analysis. Philosophy was considered fundamentally as a generalization of the results of science. Typical slogans were "science is by itself philosophy" and "philosophy overboard." The problem with this view was that it tended to become too conservative scientifically, barring the entry of new scientific ideas in conflict with the old theories. In brief, the mechanicists did not provide room for what Thomas Kuhn described as scientific revolutions.

The "dialecticians" with good reason found the mechanicist view too simpleminded. They envisaged a fundamental role for philosophy in the development of science. Deborin maintained that "materialist dialectics as a higher form of theoretical thinking can give a significant push to the development of natural science itself, now and then it can open new paths, it can help the investigators of nature to lift natural science up to a new and higher level."[10] For these philosophers the aim was a unified and "dialectical" natural science where philosophy played a leading role. The problem with the dialectical view was that it as well could be used to suppress the development of science. When two schools in science clashed, philosophy turned out to be a treacherous judge, as we can see in the debates between genetics and agrobiology that developed in the 1930s.

In a 1926 lecture at the Communist Academy,[11] Serebrovskii was already forcing a choice between Mendelism and Lamarckism. He declared that "[i]f the theory of Mendel and Morgan really cannot be brought into agreement with Marxism, then either the theory is not true or our Marxist worldview has not been worked out in some respect." Serebrovskii's answer to this dilemma was that the claims of incompatibility were due to misunderstandings of the theory. If anything, the theory of Mendel and Morgan was the result of a truly dialectical approach to nature (Serebrovskii 1926). Other angry young Marxist geneticists went further in their philosophical rejection of Lamarckism. According to one of them, Nikolai P. Dubinin (1907–1998), Lamarckism represented a "typically mechanistic" method, and its conception of hereditary change was "radically false." He was not satisfied with the prospect of a new synthetic theory built on experimental results contributed by supporters of Lamarckism as well as Morganism.[12] In Dubinin's opinion the two represented opposed worldviews and there was only one solution, namely, a struggle to the bitter end (Dubinin 1929, 88–89).

Through the victory of "dialecticians" over "mechanicists" in 1929–1930, the bond between science and official philosophical ideology was strengthened. This meant that biology, and genetics in particular, became more politicized. The next round of philosophical discussions took place in 1931. This time Deborin lost and was condemned for left-wing deviancy. But philosophical fervor remained a central element of the debates over science policy, reducing the weight of strictly scientific arguments. With the rise of Lysenkoism, ideological dogmatism and an irreconcilable attitude toward scientific opponents came back to haunt the geneticists. Dubinin remained a central figure in the ideological and political debates over genetics well into the 1980s. He defended classical genetics throughout, but often with a problematic philosophical or political twist.

As I have pointed out, the geneticists' conviction of the falsity of Lamarckism was by no means shared by all other biologists. There were good reasons to claim that its wholesale philosophical condemnation was in conflict with sound science. This view was defended, for instance, by the British geneticist J. B. S. Haldane. But in the Soviet Union serious scientific discussion was suppressed, and when Lamarckism came on the offensive with Lysenko, the impact of scientific arguments had been seriously reduced.

"CREATIVE DARWINISM"

The development of an atheistic worldview was another aspect of the Soviet-Marxist appropriation of natural science. The theory of biological evolution, with its relevance for the historical origin of man and its reputation for opposition to religious myths, became a central part of the official Soviet worldview. In 1931 the first chair (*kafedra*) of "dialectics of nature and evolutionary theory" was established at the University of Leningrad with Isaak I. Prezent as head.

Isaak Izrailevich Prezent (1902–1969), from Toropets, a town in the Pskov district, was a *vydvizhenets* (pushed up) who advanced quickly in his academic career. He became *komsomol*[13] secretary in 1920, member of the Communist Party in 1921, and political commissar in the Red Army in 1923. In 1925 he came to Leningrad, where he "studied at the university and taught at party schools and workers academies."[14] His formal education was brief. Already in 1926 he "finished" the university, according to an official biography. He then worked for some time as "senior scientific collaborator at the Leningrad branch of the Communist Academy" and was president of the Society of Marxist Biologists.[15]

Prezent was a leading activist in the "socialist reconstruction of science." Biology was a natural choice for a *vydvizhenets* science ideologue and politician with a minimum of natural scientific training. Biology contained less theoretical technicalities and specialized knowledge than physics and chemistry, and it had a more immediate ideological relevance. In an article in the Communist Party's theoretical journal, *Under the Banner of Marxism*, "The Problem of Scientific Cadres in the Light of Bourgeois Biology," Prezent (1931) picked the geneticist Iurii Filipchenko as a representative of bourgeois science and his eugenics as the point of attack. He argued that eugenic theories about inheritance of human abilities conflicted with the attempts to recruit scientists from the lower social classes. A theory that saw theoretical abilities as inherited rather than acquired was contrary to the party's political program of creating a new science with cadres of a different social origin. One of Filipchenko's interests was pedigrees showing how scholarly achievement was inherited within certain families.

In 1932 Prezent lectured to a conference of science teachers on "Class Struggle on the Natural Science Front," attacking traditional liberal ideals of science. He ridiculed the ideal of "pure science" for aiming at true knowledge independently of political loyalties and immediate practical usefulness, quoting prominent scientists such as the geochemist V. I. Vernadskii and the plant physiologist Liubimenko. Prezent's aggressive rhetoric associated the liberal ideals with the philosophical heresies of Deborin and his followers. In the "theoretical basis of the rotten-liberal attitudes" of Deborinites and others with "counterrevolutionary Trotskyist" tendencies, Prezent pointed out the contemplative attitude toward nature and the nonpartisan ideal of science as major sins. Soviet science should be practical and always on the side of the proletariat. "The leader of our party, comrade Stalin, has called the party to strengthen its vigilance on the theoretical front" (Prezent 1932, 3). The rhetoric and epistemology of Bukharin was deployed in a frontal attack on the traditional ethos of science.[16]

Prezent's teaching of evolutionary theory in the 1930s focused on Lamarck and Darwin. The inheritance of acquired characters was a central theme, and the bottom line was that one must avoid so-called mechano-Lamarckism, that is, a Lamarckism that was not dialectical. But Mendelian genetics received little attention.[17] Prezent became the foremost propagandist of Soviet "creative Darwinism." This teaching gave ample room for the inheritance of acquired characters and objected strongly to the idea of stable genes that would not change in adaptation to environmental influences. The environment did more than just select from given hereditary variation; it actively created variation in a certain direction. Prezent's course in Darwinism in the 1940s was officially described as follows:

> The theory of Darwin is presented to the students as a result of many years of analysis of practical human activity in the areas of animal and plant breeding. Particular attention is paid to the struggle for survival, the role of jumps in evolution, to determinate or indeterminate change, etc. A recurrent theme through the whole course is the creative role of natural selection. The course ends with sharp criticism of anti-Darwinism in all its different manifestations and the theoretical foundations of a new creative Darwinism.[18]

Prezent was known for his brilliant rhetoric and polemics. The kind of popular "Darwinist" doctrine that he developed had a superficial and flexible connection to scientific knowledge. It is characteristic that Prezent started as a defender of Mendelism,[19] but with the philosophical and ideological changes in the early 1930s, he soon saw other opportunities and turned to Lamarckian ideas and Lysenko's agrobiology.

THE LENIN ACADEMY OF AGRICULTURAL SCIENCE (VASKhNIL)

The Lenin Academy (VASKhNIL) was formally established in June 1929 with Nikolai Vavilov, Nikolai Gorbunov, and Nikolai Tulaikov as vice presidents. Gorbunov represented the political control of science. He was an engineer by training, had been Lenin's private secretary, and had become secretary for the council of ministers, heading the central Soviet government. From the early 1920s Gorbunov had played a central role in the organization of scientific research and development. In particular he had been active in the development of agricultural research, collaborating closely with Vavilov. Among other functions, the new academy of agricultural science took over the coordinating function of the scientific committee of the ministry of agriculture.

Immediately after the crucial decision about VASKhNIL had been taken, Gorbunov wrote to Vavilov, telling him that he would have to cancel a planned expedition. The development of agricultural research through VASKhNIL was top priority to the Soviet government, and Vavilov's participation was needed in working out the structure of the new organization. But Vavilov protested. In a long reply he explained that the research expedition was an essential part of VIR's program for investigation of the world's resources of cultivated plants. If the leading scientists were overburdened by administrative duties, they would not be able to keep up with research and the scientific level of the whole institution would suffer, argued Vavilov. He got his way. In early June he left for central Asia and the Far East, while Gorbunov took charge of organizing VASKhNIL. As the campaign of forced collectivization was set into motion, a new ministry of agriculture for the whole of the Soviet Union was also established by December 1929. Toward the end of the year Nikolai Vavilov returned from travels in the Far East to take up the presidency of VASKhNIL. Among the other members of the earlier scientific committee of the Russian Ministry of Agriculture, Nikolai Tulaikov continued to play a central role in the new organization.[20]

In contrast to biological and other natural science agricultural research, social science studies of agriculture were on the firing line from the beginning of collectivization. Already at a conference in December 1929 Stalin condemned agricultural economics for "lagging behind practice." He described how the Soviet Union was now creating a new practice, a collectivized agriculture, and new theory had to be built on this practice. "It is well known," Stalin warned, "that scientific theory, if it is a real scientific theory, provides the practitioner with power of orientation, clarity of perspective, confidence in his work, and belief in the victory of his cause" (Stalin 1929, 142). Like Bukharin in 1931, he emphasized the instrumental aspect of science and its subordination to practical political goals. The invocation of "practice" became a powerful and blinding argument in debates over science policy throughout the Stalin period.

The Lenin Academy had a scientist at its head, and much of its work revolved around a series of larger and smaller scientific meetings. These meetings were dominated by the leading scientists, and their debates over scientific and science policy issues attracted great public attention. However, the Lenin Academy had from the beginning a bureaucratic and hierarchical character quite foreign to traditional scientific academies. Its close association with the Ministry of Agriculture restricted scholarly autonomy. The so-called presidium of the academy was a secretariat of administrators rather than a committee of scholars. Minutes from the early meetings of the presidium are filled with administrative rather than scientific concerns. Often no scientists were present and the meeting was run by ministry officials alone.[21]

As president of VASKhNIL, Vavilov emphasized its central political and administratrative role in the ongoing revolution of Soviet agriculture. Collectivization of agriculture and the reorganization and expansion of agricultural research were two aspects of the same drive toward more efficient production, he proclaimed in a speech of January 1930. In Vavilov's words, VASKhNIL was to be "the academy of the general staff of the agricultural revolution"—the "general staff" being the Ministry of Agriculture. The simultaneity of "the beginning of the revolution in agriculture and the creation of an All-Union Academy of Agricultural Science is deeply significant and determines the basic tasks of our academy," he said.

Vavilov pointed to economic success as a criterion of good science. In tune with the current Soviet-Marxist theory of science, he emphasized the need for closer contact between theory and practice: scientific research should be "tightly linked to production." The scientist "must enter into the spirit of practical production," and the results of his work should be "immediately transformed into production." The demand for economic efficiency was partly to be answered by a new style of collective work, team work, like in an industrial factory. "It is necessary to establish factory-type productive conditions in order to maximally accelerate research work" (Vavilov 1930a).

A revolutionary program for agricultural science was published by the presidium of the VASKhNIL in a set of theses in February 1930. Large-scale industrial organization was the model for VASKhNIL. Thesis number two said: "small-scale farming is the basic obstacle to a successful development of scientific-experimental work in agriculture." Thesis four held that the "existing anarchy and individualism in scientific work, rooted in the capitalist conditions of production, fundamentally contradicts the planning principles needed for the building of socialism." This thesis also stressed the necessity of a "decisive fight against bourgeois and petit-bourgeois tendencies" in the attitudes, approaches, and methods of agricultural scientists.[22]

In his technological optimism Vavilov did not see to what extent collectivization disrupted the social structures that agriculture depended on. At the end of 1929 a friend, former brother-in-law, and professor of agricultural economy told Vavilov about the worrying development in the countryside over the last months. Agriculture was conservative, said this friend. It needed reform, but careful reform. The present policy threatened to create a crisis, driving away large numbers of the most capable workers. This could lead to destruction of the whole agricultural system, he argued. But Vavilov was reluctant to accept his judgment. It is not so serious, he objected; even if many peasants leave for the towns, there need not be any misery. He pointed to the United States where more than seventeen million had left the farms between 1910 and 1920, while the cultivated area and the production rose. "Science, that is the main force!" said Vavilov. The important thing is that "*sovkhozes* [state farms], *kolkhozes* [collective farms]," or whatever there may be, approach their tasks "scientifically."[23]

REVOLUTION AT THE INSTITUTE FOR PLANT INDUSTRY (VIR)

According to official figures, the number of students in higher technical education more than tripled between 1928 and 1932. In the agricultural sector the increase was below average but still amounted to more than a doubling. Though the official figures were inflated in various ways and student numbers dropped as most of the radical pedagogical reforms were abandoned in 1932–1933, the cultural revolution marked a watershed for the Soviet intelligentsia. The combined effects of educational, social, and economic changes during the first five-year plan created a revolution in the situation, attitudes, and composition of the Soviet intelligentsia.[24]

From 1930 Vavilov's Institute for Plant Industry (VIR), the largest and most important institute under VASKhNIL, had responsibility for a large program for training research specialists. Some of these "aspirants" (graduate students) were more interested in politics than in scientific studies. A vigorous criticism soon started: the work of the institute lacked contact with practical life and the research program of Vavilov and his associates was too theoretical. The criticism did not only affect student behavior. In April 1931 Vavilov complained to the presidium of VASKhNIL:

> One may disagree about principles and subject them to discussion, but unfortunately matters have gone further than this. In fact there

happens daily some kind of open or clandestine action to change parts of our working program. Only the return of the director from abroad has somewhat slowed the tempo of these events.[25]

In a letter to his deputy director at VIR in August 1932, Vavilov expressed his dissatisfaction with the excessive theorizing about science that he felt was stealing time, resources, and interest from the doing of real scientific work. The political activity also distracted the students' attention from the international cooperation that was essential to scientific progress. Vavilov was on his way to the international congress of genetics in the United States and complained that his efforts to bring other Soviet scientists to the conference had been blocked by the political authorities. "I am going alone to this important congress of genetics while we often send many people to congresses that are really of rather minor importance." As an example Vavilov mentioned the international congress on the history of science and technology that he had attended together with Bukharin the year before.[26]

One reason for the depressed mood in Vavilov's letter was his fruitless efforts to promote active Soviet participation in the international genetics congress in America and locate the next congress to Russia. In March 1932 he had written to the minister of agriculture, Iakov Iakovlev, urging that a substantial and representative delegation be sent, including senior scholars as well as young apprentices. This would demonstrate the achievements of Soviet research and show the world that the Soviet Union was well prepared to host the next International Congress of Genetics. Many scientific institutions as well as individual scholars had received invitations to the congress in Ithaca, explained Vavilov in his letter to Iakovlev. It was high time to decide who was to participate so they could properly prepare their lectures, exhibits, and so on.[27] But the political leadership was not eager to support foreign travels for Soviet geneticists, and Vavilov had to represent his country almost alone. A number of Soviet geneticists had prominent places on the program but did not show up.[28] As a result of this Sweden became first choice for the next congress.

In addition to violent internal criticism spearheaded by political activists among the research students, there were also external interventions. In early 1933, while Vavilov was still in America, a number of the leading scientists at VIR were arrested on suspicion of subversive political activities (Pavlukhin 1994, 359).[29] Some were able after a period of "work on the periphery" to return to "the center,"[30] but others, including Nikolai Maksimov, were permanently lost to VIR.

In April 1933 Vavilov described the difficult situation in VASKhNIL and its institutes in a letter to Tulaikov. The "removal" of many leading experts together with the assignment of many new tasks had created a situation where it was hard to pursue proper scientific research. A swelling central administration of VASKhNIL attempted to carry out "so-called scientific work," which was done better in the institutes. Vavilov apparently wanted more independence from the Ministry of Agriculture, but the party chief of agriculture, Iakov Iakovlev, did not agree.[31]

The negative effect of the cultural revolution on the research work was acutely felt in VIR's sections for plant breeding and plant physiology located in Detskoe Tselo, just outside Leningrad. In January 1934 Vavilov wrote an official letter to the local party committee. Because of "hypertrophy of distrust toward specialists" and general suspicion, senior scientists threatened to leave. Among the names that Vavilov mentioned this time were the physiologist V. Razumov, the cytologist Grigory Levitskii, and the geneticist Georgi Karpechenko. The "leading personnel" in Detskoe Tselo were young and had little scientific authority and experience, Vavilov pointed out. The cooperation of the senior specialists was therefore essential for successful results. With respect to the working program, Vavilov objected strongly to the idea of dropping work that was not related to the local region. Detskoe Tselo was a national center with general methodological responsibility. As an example of the narrow perspective of the young selectionists, Vavilov mentioned their disregard of work on vernalization.[32]

A few months later Vavilov tried to persuade Lysenko to come to Detskoe Tselo to help with the organization of physiological research. There is a strong wish among "our boys in Detskoe Tselo" that he should come, explained Vavilov. He suggested that Lysenko should spend at least a week two or three times a year to "help, especially the younger workers, to carry out their vernalization work better and more rapidly."[33] Vavilov felt that he needed Lysenko's help to lead the young zealots into more productive lines of research work.

REORGANIZATION OF VASKhNIL, 1934–1935

After some years of explosive growth, the unrealistic promises led to a crisis for VASKhNIL. The myth that earlier "bourgeois" science was a "pure science" divorced from practical life had helped produce inflated expectations. Vavilov, Tulaikov, and other leading scientists had held out promises that the combined application of economic planning and advanced scientific research would quickly raise agricultural yields. Research planning and a revolution in scientific method was to raise Soviet agricultural science to a new level of practical efficiency and theoretical fruitfulness. What emerged instead was crisis, both in agriculture and in science.

Both public opinion and the political leadership had great confidence in the ability of science to do wonders. When this did not happen, one naturally looked for faults in actual organization and methods of science. It was not hard for the critics to demonstrate that the ideals of unity between theory and practice had not been fulfilled. The distance between activity in basic science and practical achievements still appeared great, especially to the outsider with little scientific knowledge. While the research empire of Vavilov had produced masses of theoretical and factual knowledge in voluminous scientific treatises, the problems of practical agriculture had piled up. Other reasons for this "split" between science and practice were that state support for agriculture was much less than expected and that *kolkhoz* peasants lacked motivation to use new technologies.

Agricultural production had shown a strong and steady rise during the NEP period, but after collectivization started, the growth had diminished or turned into decline (Wheatcroft, 1984). On the then current ideological outlook, it appeared likely that in spite of all the efforts to the contrary agricultural science was still "locked up" in the academic ivory tower. Even the responsible scientific leaders such as Vavilov could not deny that there was some truth in this accusation. "Do not lag behind life!" became a constant exhortation to science from party and government in the mid-1930s. This slogan expressed a recommendation that science should learn from the socialist reconstruction of society, as well as a threat that a science incapable of learning would be eliminated.

Tension and mutual disenchantment grew between scientists on the one hand and bureaucracy and political leadership on the other. Growing opposition to the old established specialists was also found within the scientific community. Among the young generation who had received their higher education after the 1917 revolution, the revolutionary spirit was strong, and many sided with the political establishment rather than the scientific.

In May 1934 the Ministry of Agriculture issued a report on the reorganization of VASKhNIL. On receiving the report, Vavilov reacted with a telegram to the minister. The proposals were in general quite reasonable, but the criticism of earlier work was too negative. It should not be forgotten that important results had been achieved, for instance, in developing and introducing new plant varieties, argued Vavilov, and it was unreasonable "to lay the responsibility for the operative work in seed production and animal breeding on the research system."[34] Vavilov's comments presupposed a distinction between scientific and economic success. The science may be good even if it does not lead to practical economic success. There may be other reasons for economic failure. Vavilov implicitly invoked the difference between scientific research and the application of its results in practical agricultural administration.

However, the decree on VASKhNIL that the Council of Ministers issued on July 16, 1934, was sharply critical of the work done so far, on the scientific as well as the practical side. Having listened to the account of the development and achievements of VASKhNIL, delivered by its president, Vavilov, the council found that the academy "had not fulfilled the basic task which it had been assigned." Inadequate organization and narrow specialization were among the basic faults that had prevented contact with practical agriculture. The academy had not been able to provide sufficient "scientific generalization of the mass experience" of leading state and collective farms. But there were bright spots. Lysenko's vernalization and Vavilov's World Collection were mentioned as more or less equivalent positive achievements of agricultural science.

In the following year, the Lenin Academy of Agricultural Science was radically reorganized. Many research and experimental institutions were transferred to the agricultural administration centrally or locally. In 1934 the academy included 111 research institutes and close to 300 research stations of different kinds. When the process was completed in 1935, only 12 main research institutes and their branches throughout the country remained under VASKhNIL (Vavilov 1935a, 16–18).

At the same time as VASKhNIL was undergoing radical reorganization, Lysenko made a speech to the Second All-Union Congress of Collective Farm Shock Workers on February 14, 1935, a speech that demonstrated his rising political status and his change to a new and aggressive style. He started by explaining the difference between bourgeois and socialist science. Echoing Marx, Bogdanov, and Bukharin, he said that while the basic content of the first is observations and explanations of phenomena, the second aims to "alter the plant and animal world in favor of the building of a socialist society." Lysenko then went on to explain how he had developed the theory of stages. His success depended on a collective style of work. Bourgeois scientists in capitalist countries worked individually and in isolation from practical problems. It was absolutely necessary, argued Lysenko, that the "mass of *kolkhoz* workers" intervened into research work on selection and genetics. He ended rhetorically with an apology for his clumsy and nonacademic form. I am "not an orator," said Lysenko. "I am only a vernalizer." And Stalin applauded from his seat: "Bravo, comrade Lysenko, bravo!"[35]

The struggle for bigger harvests was the theme of the accompanying front-page editorial of *Pravda*, stressing once again the importance of unity between theory and practice in agriculture. Lysenko's vernalization was presented as a brilliant example of how scientific theories can be quickly transferred to practical agriculture and prove their worth. Already in the first year after the discovery, trials were made on thousands of hectares, the following year on hundreds of thousands, and now on millions. This was the right way of using "the word of science" to change agriculture, according to *Pravda*.[36]

While Lysenko's star was rising, a clear sign of Vavilov's declining favor with the government was the cancellation of the celebration of the tenth anniversary of the Institute for Plant Industry, VIR. This was also planned as a celebration of Vavilov's twenty-fifth anniversary in scientific research. From the summer of 1934 the staff of VIR had been preparing to make this double anniversary a big public manifestation. But a few days before the celebration was to take place, on February 26, 1935, the institute was suddenly informed that the event had to be postponed until after spring work in agriculture had been finished. Vavilov appealed to Iakovlev, now party boss of agriculture, and to the minister, explaining the depressing impression that the cancellation had made on the staff. It appeared as a "verdict of distrust of the enormous, in some areas literally heroic, work" of the institute, wrote Vavilov.[37]

The reorganization of VASKhNIL was concluded by a government decree in June 1935. Altogether fifty academicians were appointed, including Michurin (posthumously). Lysenko and a number of people who were bureaucrats rather than scientists, some of them without any scientific training, were also appointed. Nonscientists were placed in the two key posts as president and scientific secretary. An earlier deputy minister of agriculture, Aleksandr Muralov, replaced Vavilov as president, and A. S. Bondarenko became secretary. Muralov was an old Bolshevik who had participated in clandestine party work before the revolution and had been in the central agricultural administration since 1928.[38] To the three positions as vice presidents were appointed leading scientists, Vavilov, Meister, and M. Zavadovskii. In some ways Vavilov was probably quite relieved to step down from the presidency of VASKhNIL. The administrative responsibilities made it difficult to find time for scientific work. He had had to fight hard with administrative and political bosses to obtain leave for his beloved plant-collecting expeditions.

The new organization of VASKhNIL implied a clearer differentiation between theoretical and practical work. A large number of the most practically oriented institutes and branches were transferred to the agricultural administration. VASKhNIL retained only those parts of the system that were most oriented toward theoretical science. At the same time, its central organization became more like that of a traditional academy. The main criterion for becoming an academician or a member of the presidium was prominence as a scientific researcher. But there was also a substantial representation of agricultural administrators. Of the fifty academicians, nine were appointed by the old Academy of Sciences and forty-one by the Ministry of Agriculture. Thirty-six are listed as professors, one had a doctorate, and thirteen were without academic titles. Among the latter we find VASKhNIL officals such as Muralov and Bondarenko and the party chief of agriculture, Iakov Iakovlev.[39]

VASKhNIL had so far been run mainly by the secretarial bureau with a somewhat ad hoc use of scientific experts. Vavilov was not a very active participant in affairs outside his own field of plant breeding. Together with Gorbunov, he had laid the structure of the whole system. But it was beyond his capacity to participate actively in running it. This is the clear impression of the minutes of meetings in the presidium from 1929 to 1935.[40] With the reorganization that was concluded in June 1935 political control was tightened by giving the posts of president and academic secretary to nonscientists. A main assignment of the new leadership of VASKhNIL was to make the institutes and researchers concentrate more on pressing practical problems and not use so much time to pursue purely scientific questions.

IMPLEMENTING THE "UNITY OF THEORY AND PRACTICE"

The most important means to improve practical results was the "liquidation of the break between scientific research work and practical agriculture," said Muralov in his speech to the first meeting of the new VASKhNIL in June 1935 (Muralov 1935b, 3). Among other things he found the dissemination of research results to the public unsatisfactory. Many scientists did not write about their results in the general press, he complained (7). Some of the results were not even published in the scientific press, but remained in the scientists' drawers (4).

Vavilov agreed that the lack of an effective organizational link to the agricultural producers had been a major weakness in the old VASKhNIL. But in his opinion the most serious fault had been a lack of scientific leadership and quality. The agricultural research system lacked "an active, competent, and continuously working scientific collective." Vavilov saw that bureaucratization and the accumulation of research workers with deficient qualifications had become a serious problem in the system that he had been heading. His speech to the first meeting of the new VASKhNIL included strong criticism of a policy for which he himself had considerable responsibility. He now urged that the reorganization should aim to correct these weaknesses.

Though Vavilov was loyal to the decisions of the government, his argumentation diverged from the official interpretation of unity between theory and practice. He stressed that the key to a more useful science lay in strengthening scientific quality. "The absence of the authority of a strong collective, the opinion of which was binding and indisputable, led to hypertrophy in many of our scientific institutions," he explained. The lacking influence of competent scientific judgment produced much unnecessary and poor work. Presumably, Vavilov genuinely believed that the reorganization also included positive possibilities. For instance, the appointment of a group of prominent scientific specialists as academicians would help create the "scientific collective" that VASKhNIL had lacked so far.

Once more he also objected to the tendency to a superficial and only negative evaluation of the results that had been reached, for instance, in plant breeding (Vavilov 1935a, 18–20).

Muralov's concluding assessment at the first meeting of the new VASKhNIL gave special emphasis to the principles of the ongoing scientific revolution. He said that the "fight for the party line in Soviet agricultural science is the most important task" of VASKhNIL. And he described the rift between theory and practice as an "oppressive inheritance from bourgeois science, with its narrow perspective, with its fear of the working masses, with its service for kulaks and landed proprietors." This rift was also "a capitalist remnant in the mind of the individual scientific workers," which made them "lock themselves up in their laboratories and not move further than their greenhouses." The urging of the minister of agriculture to use the state and collective farms as their experimental fields, to conduct practical experiments on a grand scale instead of puttering alone in the laboratory, was the key to overcoming the split between science and production, according to Muralov (1935b, 3–4). He repeated the main points in *Izvestia* on June 29 under the title "Don't Hang Back from Life."

The demand that not only the results but also the research itself should be brought out to the peasants suited Lysenko excellently. He was already famous as the standard example of how agricultural scientific research ought to be conducted under socialism, in close contact with the peasants. In May 1935 the Academy of Sciences organized a meeting on agricultural research where Muralov pointed to Lysenko's theory of stages in the development of plants as a result of true socialist method in science. This theory had been applied in the vernalization of seed grain, and it was "carried forward by millions of peasant hands, to come true under productive conditions in socialist fields, giving a considerable rise in yields" (Muralov 1935a, 13).

If one accepted the doctrine of unity between theory and practice, believed in the practice criterion of truth, shared the regime's aims of social and economic progress, and took Lysenko's claims about increased yields to be correct, it was hard to escape the conclusions of Muralov. More than any other scientist, Lysenko appeared as the one who had succeeded in practicing the new ideals of scientific research. Muralov merely pointed out the consequences of the methodological revolution that Soviet science had supposedly gone through.

The Ministry of Agriculture did not behave particularly narrow-mindedly and fanatically in following up the new principles for scientific research. Lysenko was their mascot and received much support, but the expertise of the old specialists was respected. As a young and relatively inexperienced member of the scientific community, Lysenko at this point became no more than an ordinary academician of VASKhNIL. The three positions as vice presidents were reserved for more experienced scientists. All three would later stand up as sharp critics of Lysenko's genetics and breeding methods, first Meister and Zavadovskii, and then Vavilov.

"SOCIALIST COMPETITION" IN SCIENCE

In industry "socialist competition" had been widely introduced to increase productivity. Stakhanov was hailed as a hero worker producing several times the average output. In 1936 VASKhNIL adopted a program for "socialist competition" in science, in accordance with the idea of unity between theory and practice. At the end of December 1935 Lysenko spoke to a public propaganda meeting of agricultural workers. The front page of *Pravda* showed Lysenko on the rostrum with Stalin and other top leaders on the podium. Lysenko was met with standing applause, and his speech was presented over a whole page. Vavilov, who was also speaking, shared a page with three others and received only modest applause.[41]

In cooperation with Jakovlev, Lysenko adopted a threatening tone. He spoke about those scientists who opposed the principles of his new methods in plant breeding and worked against their practical application. When Iakovlev asked for names, Lysenko mentioned Vavilov and some leading plant geneticists.[42] He also quoted what Stalin had said to a congress of Stakhanov shock workers one month earlier, in November 1935:

> If science was as it is being described by some conservative comrades it would have been lost to humanity long ago. Science is called science precisely because it does not recognize fetishes, does not fear to lift its fist against the obsolete, and closely listens to the voice of experience, practice.

Especially the last sentence was often repeated in debates over science policy in the following months and years.

Lysenko used the occasion to challenge his scientific opponents to participate in a "socialist competition." Let us "stop abstract discussions and turn toward the only true proof—toward practice," he said. If an approach does not produce a solution, "I shall be the first to change to new working methods." Lysenko proposed that the principles of the Stakhanov movement be introduced in scientific research, setting precise goals and norms for the output. Why should science be without concrete and precise goals and norms when they had been introduced in the practical world?[43]

The growing conflict over the principles of genetics and breeding was becoming troublesome for the agricultural administration and VASKhNIL. It was embarrassing that scientists appeared to use their resources for infighting instead of helping agriculture. Science that should be the great clarifying and progressive force was threatened by impotence because of internal strife. The impatience with scientists' disagreement and inability to give clear and final answers was increased by the difficult economic situation. It was becoming more and more clear that agriculture was lagging behind in the great process of economic modernization. Intense public campaigns to increase agricultural production were launched in the mid-1930s, focusing on grain production as the key factor. But 1936 produced the poorest grain harvest in many years (Wheatcroft 1984).

The idea of socialist competition in agricultural research was eagerly pushed by VASKhNIL. This appeared a good way to harness science to the campaign for increased agricultural production. It applied industrial principles of organization and planning to scientific research. The first issue of the VASKhNIL bulletin for 1936 carried a leading article, "The Stakhanov Movement in Science," based on Stalin's November 1935 speech to Stakhanov shock workers. Every scientific research institution within the VASKhNIL system was already required to have a production plan. But the political and adminstrative leaders of VASKhNIL could see no enthusiastic effort to fulfill the plans. One reason seemed clear to them: the atmosphere of "marketplace" and "chaos" characteristic of scientific institutions in the old system had not yet been eliminated (Bondarenko 1936).

The scientific workers themselves had not been able to reach agreement on a criterion of success, observed Bondarenko. Some held it to be the publication of scientific books and articles; others thought the main result should be instruction in new production methods to those who carried out the practical work. The answer to this stalemate, according to Bondarenko, was to apply the practice criterion in a Stakhanovite spirit. It is "completely clear that the success of our work will only gain full weight when the application of the new methods gives a large effect." The article was illustrated with the portraits of four "Stakhanov" scientists who had just received the Lenin Prize, Lysenko, Meister, Iogann Eikhfel'd, and N. V. Tsitsin. (Eikhfel'd was a pioneer in pushing agriculture into the far north with the use of vernalization, and Tsitsin was trying to produce a perennial wheat by crossing with couch grass.) Bondarenko concluded by promising 1936 to become a year of "decisive restructuring of our scientific work" (Bondarenko 1936, 3).

Toward the end of April 1936, the scientific council at the Institute for Selection and Genetics in Odessa, which Lysenko now directed, published an appeal for "competition between workers in agricultural science." By focusing attention on "concrete methods to increase the yield" one could avoid fruitless scholastic debates, declared the council. The various groups of research workers would commit themselves to accomplish specific tasks within a set time, and the results would be subject to public evaluation. In this way the practice criterion of truth was to be made operational. The ability of the different groups to create new plant varieties and techniques of cultivation would demonstrate the truth or falsity of their theories.

The declaration listed tasks taken on by the Odessa institute. For each task the link to Lysenko's theory of stages was noted. But it was genetic and not physiological aspects of that theory that were emphasized. Most of the tasks were in the breeding of cereals, one in cotton breeding. The competition was open to all research institutions concerned with plant breeding, but Vavilov's Institute for Plant Industry in Leningrad and Meister's station for plant breeding in Saratov were specially challenged. VASKhNIL together with *Pravda* and the daily agricultural newspaper *Sotsialisticheskoie Zemledelie* (Socialist Agriculture) were proposed as judges.[44]

During the following months "socialist competition" in agricultural science was a central topic in the VASKhNIL bulletin, with regular references to Lysenko and his new "style" of scientific work. The constant urge was that scientific institutions and workers should "study and generalize" the "grandiose successes of the Stakhanovites on *kolkhoz* fields and farms."[45] In the beginning of August 1936 there was a meeting of the directors of the research institutes of VASKhNIL with the leader of the scientific section of the Central Committee and his deputy. The two were worried about the reluctance of the institutes to take on the new style of work. There were many good words but meager results, said the deputy. Institute directors tended to take a "formal" and passive attitude, and self-criticism was weak (Doroshev 1936).

Muralov repeated the call for socialist competition in agricultural science at an October 1936 meeting of VASKhNIL. He praised the concrete character of the tasks the Odessa institute had set itself, "each of them included not only the working through of this or that theoretical problem, but also—and that is the most important—the transfer of the results ... into production ..." (Muralov 1936b, 1). Muralov noted slow reactions and lacking enthusiasm in most research institutes, but a number of bilateral competitions were now in progress, among them Vavilov's VIR against Lysenko's Odessa institute (Muralov 1936b, 2). Within a month all institutes must have such agreements of mutual testing, declared Muralov. In industry and other parts of the economy, production plans were regularly surpassed under this system, so why should not agricultural science benefit as well (Muralov 1936b, 5)? "Socialist Competition— to the Highest Level," was the title of Muralov's address. In view of the new constitution soon to be discussed in the national assembly, it was time to intensify socialist competition in agricultural science.

Muralov's address was printed in the December issue of *Selection and Seed Growing*, the monthly "scientific-industrial" journal of the Ministry of Agriculture, in preparation for the fateful upcoming December 1936 conference on genetics and breeding. The Ministry of Agriculture and the political leadership of VASKhNIL no doubt saw this conference as a means to highlight and intensify socialist competition in agricultural research. In the same issue, next to Muralov's, was an article by Lysenko, "The Test of Practice" (Lysenko 1936b). The specific tasks that the Odessa institute had taken on in April had all been solved. But it had also been promised that these achievements would entail further development of the basic theory and produce new practical results. Now Lysenko announced a sensational new practical result, namely, that in the spring of the following year, 1937, a spring variety of the winter wheat Kooperatorka, one of the best winter wheats, would be ready for sowing. The silent assumption was that this new variety would retain the high yield and other valuable qualities of the parent.

In essence, Lysenko's challenge was to produce a much-needed new variety of wheat by means of a theory in deep contradiction with the principles of classical genetics. He claimed already to have shown experimentally that it was possible to produce directed mutations by selection of modifications, the same kind of idea that Keller had argued for and Zhebrak had experimented on.[46] Now he wanted to prove his genetic theory by putting it through a test in accordance with the practice criterion of truth. With a new, superior spring wheat produced in this manner, he would force the plant breeders and geneticists to make the same kind of retreat and recognition of Lysenko's results as Maksimov had made from 1929 on.

Most likely, Lysenko was sincere in his beliefs both in a Lamarckian type of inheritance and in the practice criterion as a valid way of deciding controversies in scientific theory. The Bolshevik politicians and administrators no doubt had a prejudice in favor of a proletarian scientist like Lysenko, and perhaps also in favor of an environmentalist thesis such as the inheritance of acquired characters. However, as good bureaucrats they also respected science and hard facts. The bosses did not venture to make scientific judgments themselves. It was the scientists' job to discuss and decide on scientific matters. But the political bosses were eager to help science become more efficient in making such decisions and to implement scientific decisions once they had been made.

Marxist theory of science provided the basic ideas on which the Soviet doctrines of science policy were developed. After a relatively pluralistic phase under the New Economic Policy (NEP) of the 1920s came a strong concentration on specific approaches. First there was a romantic wave, as expressed by Stalin in various public speeches in 1929 and by Bukharin and Vavilov at the 1931 international history of science congress in London. Many scientists were also enthusiastic because of the high status of science and the generous resources for research that followed. When the visions of an agricultural revolution born from agricultural sciences started to wilt in 1934–1935, the Soviet-Marxist doctrines of science policy revealed their more hard-nosed operational aspects. The political situation demanded quick practical results from agricultural science, and the government believed it had the right doctrine to make the results happen. Leading scientific experts saw that this turn of events leveled the way for unrealistic promises and sham results like Lysenko's, not concrete practical results as quickly as possible. But they were caught by their own earlier optimism and confined by the narrow limits of the now-established "Stalinist" discourse on science policy; the ways of effective protest and criticism had become severely restricted. The systematic terror against people with diverging views on issues with political implications scared people from speaking out on scientific matters. And even the brave few who ventured, for instance, to criticize the scientific claims of Lysenko, tended to be met by philosophical and policy doctrines rather than by scientific argument. For the Bolshevik faithful this made good sense in the light of the general "scientific" theory of Marxism that they believed in. And many scientists followed suit, one way or another. In the next two chapters, 5 and 6, we shall see how science policy influenced developments in plant science, physiology, breeding, and genetics, in the first half of the 1930s.

NOTES

1. For details about Bogdanov and his philosophy see, for instance, Grille (1966).
2. A. A. Bogdanov, ed. (1909). This collection also contained essays by the writer Maxim Gorkii and by the later commissar (minister) of culture A. V. Lunacharskii.
3. The philosophical debate among Soviet Marxists in the 1920s and 1930s is a big and complex theme that there is no room to discuss here. Let me only say the following to make my perspective clearer: In his account of the epistemological and metaphysical controversies in Soviet philosophy, Joravsky (1961, 91ff) uses "mechanism" as the name of empiricist trends, like some members of these trends did themselves. We must not let this terminology cover up the fact that a main line of conflict was between a traditional scientific realism and early-twentieth-century antimetaphysical empiricism. The effect of Stalinism in philosophy was to suppress the debate between realist and empiricist views of natural science and thus to bury the conflict rather than to resolve it. In Western philosophy of science after the 1960s the term "positivism" has often been used in a similar way covering up the difference between a traditional realist metaphysics of science and empiricist antimetaphysical views.
4. Emphasis in the original.
5. The Russian term is *zamknutyi*, which means "locked up."
6. This is Gustav Wetter's (1958) translation of the second thesis. The German original is as follows: "Die Frage, ob dem Menschlichen Denken gegenstädliche Warheit zukomme—is keine Frage der Theorie, sondern eine *praktische* Frage. In der Praxis muss der Mensch die Wahrheit, d.h. Wirklichkeit und Macht, Diesseitigkeits seines denkens beweisen. Der Streit über die Wirklichkeit oder Nichtwirklichkeit des denkens, das von der Praxis isoliert ist—ist eine rein *scholastische* Frage" (509).

"Die Philosophen haben die Welt nur Verschieden *interpretiert*; es kommt darauf an, sie zu verändern."

7. Attempts have been made to extricate Marx from his conflation of science and politics by arguing that his problem was not the Kantian "thing in itself." He took for granted both the existence of an independent external world of objects and the ability of human science to achieve knowledge about it, according to one philosopher and critic of Soviet Marxism. The idealist tendency to conflate theory with reality entered common philosophical discourse only with the neo-Kantians of the late

nineteenth century, he argues. Marx was simply worried that human knowledge and the ordinary science practiced in his time was not sufficiently powerful and "this-sided" to succeed in transforming the world. In the Hegelian terminology of Marx *gegenständliche Wahrheit* means "truth concerning objects" rather than "objective truth." And *Wirklichkeit oder Nichtwirklichkeit des Denkens* does not mean "objectivity or nonobjectivity of thought" but rather "effectivness or noneffectiveness of thought." (See N. Lobkowicz, "Is the Soviet Notion of Practice Marxian?" *Studies of Soviet Thought* 6 [March 1966]: 25–36.) There is probably a sound historical basis for this attempt to save Marx, but it was lost by the philosophical tradition. In standard English translations, for instance, it is hard to recover the benevolent interpretation. The translators seem to concur with Bukharin.

8. See for instance Gaissinovich (1980, 1988).

9. See Kolchinskii (1991) for a brief and incisive account of this development.

10. A. M. Deborin, "Engel's i dialektika v biologii" (Engels and dialectics in biology), *PZM* 1 (1926): 14–15. Quoted from Kolchinskii (1991, 41).

11. The Communist Academy was developed by the Soviet government in the 1920s as an ideologically dependable alternative to the Academy of Sciences.

12. This approach was suggested by the embryologist and later active opponent of Lysenko, Mikhail M. Zavadovskii.

13. Young communist league.

14. Obituary in *Vestnik Sel'skokhoziastvennaia Nauka* 3 (1969): 151.

15. "Akademik Isai Izrailevich Prezent," *Vestnik Leningradskogo Universiteta* 10 (1948).

16. A decade later the American sociologist Robert Merton published his classical formulation of "the ethos of science" (Merton 1942).

17. See, for instance, the collection of excerpts from classical biological authors, edited by Prezent, *Khrestomatiia po Evoliutsionnomu Ucheniu* (Reader in the theory of evolution), (Leningrad, 1935).

18. "Akademik Isai Izrailevich Prezent," pp. 100–101.

19. See "Nauchnaia khronika," *Priroda* 9 (1930): 926–27.

20. Esakov (1971, 219–56) gives a detailed account of how VASKhNIL was established.

21. TsGAE, f. 8390 (VASKhNIL archive), op. 1, ed. 1, l. 5–30. Minutes of meetings of the presidium in early 1929.

22. "Nauku na sluzhbe sotsialisticheskomu zemledelie" (Science in the service of socialist agriculture), *SZ*, February 27, 1930.

23. This story is told by M. Popovskii on the basis of an interview with N. P. Makarov, Vavilov's former brother-in-law, in 1967. Popovskii (1983, 38–39).

24. See, for instance, Bailes (1978).

25. Letter from N. I. Vavilov to the presidium of VASKhNIL, April 1931, *VL* 2, pp. 114–15.

26. Letter from N. I. Vavilov to N. V. Kovalev, August 9, 1932, *VL* 2, pp. 179–80.

27. Letter from N. I. Vavilov to Ia. A. Iakovlev, March 23, 1932, *VL* 2, pp. 163–64.

28. Among those who did not turn up were G. D. Karpechenko, M. S. Navashin, G. A. Levitskii, A. S. Serebrovskii, N. P. Dubinin, and V. Pisarev. The only prominent Soviet geneticist beside Vavilov who did register at Ithaca was A. A. Sapegin. Altogether three Russians registered. The plenary session of the congress elected a "permanent international committee" of fifteen persons representing fifteen countries. Among the members were R. Goldschmidt for Germany, Harry Federley for Finland, J. B. S. Haldane for Great Britain, H. Nilsson-Ehle for Sweden, N. Vavilov for the Soviet Union, and R. A. Emerson for the United States. O. L. Mohr from Norway was asked to serve as "temporary chairman who might initiate correspondence within the committee since it appears unlikely that the committee will be able to meet. The chief functions of the committee are: to represent the international congress until the council of the next congress is formed and to designate the country in which the next congress is to be held." (*Proceedings of the Sixth International Congress of Genetics, Ithaca, New York, 1932*, vol. 1, ed. D. F. Jones [New York: Brooklyn Botanic Garden, 1933] p. 17.)

29. A letter from Vavilov to N. V. Kovalev of August 9, 1932, indicates that Maksimov is still head of plant physiology at VIR, and that Vavilov expects no change (*VL* 2, p. 180). After his return from a trip to North and South America lasting approximately seven months from early August 1932 to February 26, 1933, he wrote to A. A. Sapegin, April 6, 1933: "amazing events" are heaping up, "20 people have disappeared from the staff" (*vybylo 20 chelovek iz stroia*), among them G. A. Levitskii, Maksimov, and Pisarev. (*VL* 2, p. 191.)

30. Letter from N. I. Vavilov to D. L. Rudzinskii, September 7, 1934, *VL* 2, pp. 241–42.

31. Letter from N. I. Vavilov to N. M. Tulaikov, April 28, 1933, *VL* 2, pp. 195–96.

32. Letter from N. I. Vavilov to the party committee at Detskoe Tselo, January 31, 1934, *VL* 2, pp. 217-18.

33. Letter from N. I. Vavilov to T. D. Lysenko, May 26, 1934, *VL* 2, pp. 233-34.

34. Telegram from Vavilov to Chernov, May 10, 1934, *VL* 2, pp. 230-31.

35. T. Lysenko, "Vernalization—A Mighty Instrument for Increasing Yields," *Pravda*, February 15, 1935, p. 2.

36. "Na bor'bu za vysokii urozhai," editorial, *Pravda*, February 15, 1935, p. 1.

37. Letter from N. Vavilov to Ia. Iakovlev, February 28, 1935, *VL* 2, pp. 274-76.

38. "Aleksandr I. Muralov (1886-1937)," *Sovetskaia Rossia*, June 21, 1966, p. 4.

39. *BV* 4 (1935): 11.

40. TsGAE, f. 8390, o. 1; TsGAND, f. 318, o. 1-7.

41. *Pravda*, January 2, 1936, pp.1-3.

42. Lysenko said: "Prof. Karpechenko, prof. Lepin, prof. Zhebrak, in general the majority of geneticists do not agree with our position."

43. "Rech' akademika T. D. Lysenko," *Pravda*, January 2, 1936, p. 3.

44. "Sorevnovanie rabotnikov sel'skokhoziaistvennoi nauki" (Competition of workers in agricultural science), *Pravda*, April 26, 1936, p. 3.

45. Editorial in *BV* 6 (1936): 1-4. Since there were twelve issues per year, this would be the June issue.

46. The ideas and work of Keller and Zhebrak will be further discussed in chapter 6 under "Dialectical Materialism in Genetics."

Chapter 5.

Vernalization

"Vernalization" was the trademark that made Lysenko internationally famous. It was a translation into English of the Russian term *iarovizatsiia*.[1] In Ukrainian *iar* means spring and *iarovoi khleb* in Russian means spring grain. "Vernalization" was in its origin a term for the transformation of winter grain into spring grain. It is still a standard scientific term for the influence of low temperatures on plant development.[2] The continued use of this term is an indication of the international scientific impact that Lysenko's work has had.

Vernalization as a measure in practical agriculture must be distinguished from vernalization as a topic of scientific research. Lysenko claimed achievements in both fields, and they have to be evaluated on different criteria. In the first case, the frame of reference is the development of efficient agricultural techniques and in the other it is contributions to theoretical knowledge about plant development. Lysenko was not the only one who underestimated the difference between the two kinds of activity and achievement, arguing that the one more or less directly implied the other. The Soviet ideology of science that insisted on the unity of theory and practice had a tendency to obscure the difference between theoretical and applied scientific research and to confuse the criteria of their evaluation. This chapter will trace the development of the debates over vernalization as both an agricultural technique and a field of scientific research. To illuminate the dynamics of these debates and show how they influenced research policy, I will present details from some representative discussions.

LYSENKO'S SUCCESS IN THE MASS MEDIA

A practical demonstration on the farm of his father in the summer of 1929 provided Lysenko with the springboard to national fame. The older Lysenko had sown a field with winter grain treated according to the theory of his son. The field gave a good harvest and sparked the enthusiasm of Ukrainian agricultural officials.

On July 21, 1929, *Pravda* announced "The Discovery of Agronomist Lysenko." A specially assigned commission from the Ukrainian Ministry of Agriculture had visited the Lysenko farm a week before to inspect its spring-sown field of winter wheat. In the judgment of the commission, the yield would be approximately three tons per hectare, while ordinary fields of spring wheat in the same region gave one ton per hectare. A representative of the ministry declared that this discovery would play "an enormous role," especially in the zone of steppe that was regularly subject to climatic conditions that killed ordinary winter wheat.[3] On July 19 Lysenko spoke to a meeting of the collegium of the Ukrainian ministry of agriculture, and on the twenty-seventh he spoke again in Kharkov, together with his father, to a meeting of scientists and agricultural administrators.[4]

Another national newspaper reported that the Lysenko experiment at Poltava had given a exceptional harvest of approximately 2.7 tons per hectare. Two rather different possible practical applications of Lysenko's method were pointed out. One was to replace autumn sowings killed during the winter by spring sowings of some vernalized variety of winter grain. But the other was to treat ordinary spring grains with the same method. This could speed up development and thus save the plants from summer drought. The report also reminded that a precise practical method still had to be worked out; in particular, a practical method to chill the moistened seed was needed.

An agricultural official said to the newspaper that it was too early for a definitive evaluation of the method, but he thought "the very possibility of feeling absolutely *insured against frost killing of the winter grain*" had "huge practical interest." The director of the Poltava experimental station wanted the Ministry of Agriculture to announce a public competition on the best method for chilling the seed. A professor and director of another experimental station used the occasion to point out their lack of equipment. The experimental stations simply could not follow up with the necessary experimental work without much better refrigerating equipment.[5]

In early October the Ukrainian minister of agriculture published an article in *Pravda* titled "The Sowing of Winter Cultures in the Spring," praising Lysenko's discovery in strong words. Most important was that it had become possible to "convert winter cultures into spring cultures" as a method to avoid the destruction of winter grain by harsh winters. Another use was to make spring grain more drought resistant. The great potential benefit to breeding work was also mentioned. On this background he declared that research to develop this method would get public support, in particular the work of Lysenko himself. The Ukrainian ministry of agriculture was also planning a large-scale practical test. One thousand hectares was to be sown in the spring with winter wheat according to Lysenko's method.[6]

Lysenko was engaged to organize this work. He moved from Gandzha to the Ukrainian Institute for Selection and Genetics (later the All-Union Institute for Genetics and Plant Breeding) in Odessa. This was one of the major centers of agricultural plant research in the Soviet Union. Lysenko became head of a new laboratory of *iarovizatsiia* (vernalization). In addition to managing the practical tests, he quickly built up a research team working on the effects of vernalization applied to a wide variety of cultivated plants. "Vernalization" soon was used in a broad sense by Lysenko's team, covering a wide variety of factors affecting the phasic development of plants. They "vernalized" with cold, heat, darkness, light, and so on. In other words, Lysenko established a research program based on his own theory about plant development.

RECOGNITION AND CRITICISM FROM THE LEADING SCIENTIFIC EXPERTS

Already on September 1, 1929, Lysenko gave a lecture at Vavilov's Institute for Plant Industry (VIR) in Leningrad.[7] In the audience were Nikolai Maksimov (head of plant physiology), Pisarev, and Prof. Viktor V. Talanov (1871–1936), a collaborator of VIR and major specialist on cereal systematics and selection. Vavilov himself was traveling in East Asia.[8]

In this lecture Lysenko elaborated on his theory that the difference between spring and winter grains is only one of degree and that there is a continuum in the effect of the cold treatment. He described an experiment carried out in the spring and summer of 1929 with cold treatment of seed for differing lengths of time from zero to seventy-five days. His method of chilling was simply to bury the grain in snow. From this and earlier experiments Lysenko gave estimates of the cold requirement of various varieties. For instance, the spring wheat Markis had "zero winterness" (*ozimost*). Cold treatment of any length would not speed up its development. The winter wheat Kooperatorka, on the other hand, needed eighteen days at zero degrees to change into a spring form. Another winter wheat, the Ukrainka, needed as many as fifty-one days.

In the discussion Maksimov was much more positive toward Lysenko's work than he had been at the big congress in January. In particular, Maksimov was impressed by its quantitative aspects. He acknowledged the "great service" of Lysenko in finding a simple method for the quantitative determination of the "winterness" of plants, expressed in days. "Priority in this discovery fully belongs to the speaker," he said. Talanov proposed that Lysenko should be asked to write up his results for publication in the series of the institute. On the suggestion of Talanov and the motion of Pisarev, who presided at the meeting, Lysenko was invited to participate in the institute's network for testing grain varieties with respect to winter hardiness and cold requirements.[9] This positive reception at VIR represented recognition of methodological and theoretical significance rather than practical value, and Lysenko was invited to cooperate with VIR in further research.

Opinions about the immediate practicability of Lysenko's method differed, on both the scientific and the political sides. At first the central Soviet government, including the Russian Ministry of Agriculture, was apparently more reserved than officials in the Ukraine, perhaps because its scientific advisors were skeptical. The agricultural daily newspaper did not bring any accounts of Lysenko's discovery until October.[10] This was a relatively sober report setting Lysenko's discovery into a wider context of research where Maksimov figured as the leading expert. In November this paper carried a series of articles by leading experts evaluating Lysenko's method.[11] Some of them discussed mainly the practical usefulness of Lysenko's method, while others saw it more as a topic for further scientific investigation. The low-temperature effect was a well-known phenomenon, and many kinds of practical application had been attempted in the past, the experts warned. Further testing was necessary to prove the practical value of Lysenko's specific ideas, and they mentioned a number of obstacles. But most of the experts also emphasized the interesting possibilities and promises that lay in the further development of Lysenko's ideas. There was considerable enthusiasm about these prospects, especially in Maksimov's article.

The first article was the most critical. It was written by the famous cereal breeder Prof. Piotr I. Lisitsyn (1877–1948) of the Timiriazev agricultural academy. Lysenko's (or his father's) merit was to have carried out a practical application of well-known principles of plant physiology. In Lisitsyn's judgement the method was suitable only for small-scale and nonmechanized farming. The large-scale farming now being developed in the Soviet Union would need quite different methods and machines both for the treatment of the seed and for the sowing. Lysenko's idea had been "coldly received," wrote Lisitsyn.

Nikolai Tulaikov was also quite critical, at least with respect to the idea of sowing winter wheat in the spring. In the dry climate of the southeastern steppe regions, the main advantage of ordinary winter wheat was that the root system was developed during the preceding autumn. The plant was thus in a good position to make use of the spring moisture in the soil. With spring sowing, the winter wheat would lose this advantage over ordinary spring wheat. The answer to present problems lay in the development of more hardy varieties of winter wheat, which were well under way. With respect to yields, winter wheat sown in the autumn would always be superior to any spring sowing, concluded Tulaikov.

Andrei A. Sapegin (1883–1946) was a botanist and plant breeder, professor at the university in Odessa, and a founder of the Ukrainian Institute for Genetics and Selection, where Lysenko was now to work. He gave a short summary of Lysenko's theory, stressing its interesting theoretical possibilities and the importance of further research with respect to its practical implications.

Maksimov's response was published some days after those of Lisitsyn, Tulaikov, and Sapegin. He acknowledged that Lysenko had made significant contributions to the knowledge of winter annual plants. In particular, Lysenko had established "the very important fact" that different varieties need different doses of low temperature to set flowers. There was, however, nothing principally new in Lysenko's work, no "discovery" in that sense. His main contribution was in the application of theoretical science to practical problems. "Both Gassner and I," wrote Maksimov, "being plant physiologists, went no further than laboratory experiments. Cold germination appeared to us to be too complicated for direct application in field farming." Lysenko had simplified the method to such an extent that it was now accessible to ordinary peasant farmers. It was "certainly impossible not to acknowledge this as a great achievement," conceded Maksimov.[12]

Not only the careful academic style with double negation distinguished Maksimov from agricultural scientists Lisitsyn and Tulaikov. He also had a tendency to praise in general but vague terms the "practical" achievements of Lysenko. But the practical usefulness of the method was precisely what Lisitsyn and Tulaikov doubted. To Maksimov a method was "practical" simply if it worked under the conditions of ordinary agricultural production. To Lisitsyn and Tulaikov the crucial thing was whether it was *more efficient* than alternatives in terms of output relative to input of labor and other resources. This illustrates Maksimov's considerable distance from the problems of agricultural practice. He was at this stage primarily an academic interested in practical applications but with little insight into how mere technical possibilities could be developed into economically competitive techniques.

An editorial comment echoed Maksimov's positive evaluations, not recognizing that he was a plant physiologist rather than an agricultural scientist: the general possiblity of converting winter into spring grain with low temperature was already well known. Lysenko's important contribution was discovering a practical method for doing it.[13]

A couple of weeks later both Lysenko and the Ukrainian Ministry of Agriculture responded to the critical comments of the experts. The scope and the theoretical tone of Lysenko's reply is revealing. He overlooked the agricultural experts and used his whole reply to argue against Maksimov. Lysenko insisted that his theory of "winterness" (*ozimost*) was quite different from Gassner's and Maksimov's theory of "cold germination." His research in Gandzha had been aimed at "a theoretical problem—what kind of phenomenon is 'winterness,' why does winter varieties not set spikes when sown in the spring?" It was only later that the theory of "winterness," which he had worked out at Gandzha, had proved its practical applicability through the experiment of his father, explained Lysenko. Much work still remained to develop a fully practical method for the vernalization of winter grains. He also maintained that the vernalization of spring grains, speeding up their development, might become a useful practical measure. He hoped that in the World Collection at VIR it would be possible to find the right varieties for such procedures.[14] Lysenko did not claim to have in hand the solution to the problem of winter destruction in the grain fields, but only to have promising ideas that ought to be worked out and tried. His reference to the exploitation of the World Collection of cultivated plants is also significant. Here Lysenko had a common interest with Nikolai Vavilov.

The Ukrainian Ministry of Agriculture disliked in particular Lisitsyn's comment that the "majority" had been negative or not interested in the discovery of Lysenko. As far as the ministry knew, "the work of Lysenko had been met with great interest" in agricultural circles, including experimenters, and especially in the Ukraine. Among the responding experts, Sapegin was the one who had demonstrated the best understanding of Lysenko's ideas, claimed the ministry.[15]

It is significant for the development of Lysenkoism that journalists and administrators in agriculture were mostly closer to Maksimov's academic perspective than to the practical agricultural orientation of experts such as Lisitsyn and Tulaikov. Neither agricultural officials nor academic scientists saw clearly enough the crucial difference between a technique that can be made to work and one that it pays to use.

PROMOTING THE VERNALIZATION OF SEED GRAIN

Lysenko's research group at the Odessa Institute for Selection and Genetics worked on two fronts. Field trials by peasants under scientific supervision were to determine yields under varying conditions. Experiments at the institute developed the general theoretical basis and new techniques of vernalization for large-scale farming. Choosing the most suitable varieties for vernalization was another main task.

In a short newspaper report of July 1930, Lysenko announced that a number of technical problems had been solved. For instance, it had been shown that the seed could be dried after vernalization. This made it possible to carry out the vernalization centrally and distribute the seed afterward to the different growers, and ordinary sowing machines could be used. As long as the water content of the seed was below 25 to 30 percent it would not be spoiled by molds. In this way fifteen hundred kilograms of vernalized winter wheat had been sent by post in parcels weighing from eleven to twenty-four kilograms to a hundred different places for sowing, not only in the Ukraine but also in other republics of the Soviet Union.

It had also been shown that vernalization would be effective even if the seed were not moist enough to start germinating. This made the process much more robust with respect to temperature regulation. Less moisture meant less danger of germinating too far and less exposure to destructive fungi. It was still too early to say anything about the yields, but Lysenko reported that for a number of spring varieties, plant development had been quickened by about one week.[16] His assessment of the robustness and general applicability of his newly developed methods turned out to be too optimistic, however. Vulnerability of the seed to excessive growth and destructive fungi remained a serious problem.

The first year of trials and experiments told little about the efficiency of vernalization with respect to yields. The only clear effect was in speeding up the development of the plants. On February 20, 1931, Lysenko lectured to a meeting at VASKhNIL. The agricultural newspaper presented his results under an optimistic heading: "Spring Sowing of Winter Cereals according to the Method of Agronomist Lysenko Has Proved Itself." But the text admitted that it was not possible to "give a precise assessment of the harvest."

Lysenko's lecture revealed that the practical testing of vernalization in regional trials had run into a series of technical difficulties. Partly because of unfavorable climatic conditions, the vernalization procedure turned out to be too complex and difficult to master for peasants under primitive conditions. This time the winter was too mild. There was a lack of snow to chill the seed grain, which therefore tended to sprout too much or be spoiled in other ways. The result was in many cases an insufficient density of plants in the fields. When the density of plants becomes too low, the yield will decrease. The result was large variations in the harvest. It also became clear that Kooperatorka was not the most suitable winter variety in all cases. It ripened too late and was therefore hard hit by summer drought. Lysenko concluded that one would have to look for other, more suitable varieties. The Kooperatorka gave very good yields in some cases, which supported the earlier promises, he argued. He did not say that even in these cases the lack of parallel sowing of spring wheat did not permit a precise evaluation of the method.

Thus Lysenko was unable to present any data that could directly support a claim to increased yields by vernalization of the seed of winter varieties. However, he did present data to show that the development of some varieties of spring wheat could be sped up by as much as thirty-five days. The real practical usefulness of vernalizing seed grain was still an open question, much as it had been a year earlier, and Lysenko's general conclusion was modest: "one already has the right to speak about the possible perspective of practical application of the method of vernalization."[17]

The meeting found, nevertheless, that Lysenko's experiments were highly promising and deserved special attention. A resolution was passed in the name of VASKhNIL supporting Lysenko's various projects and demanding the cooperation of other institutes in his research program. Cooperation with VIR and exploitation of the World Collection was specially mentioned. However, this was a meeting of "the bureau of the presidium," not of the presidium proper. Neither Vavilov, Tulaikov, nor any other of the scientific experts on breeding or growing of cereals was present.[18]

In May 1931 Lysenko and Maksimov both applied to VASKhNIL to get support for research on vernalization. Maksimov wanted to expand the ongoing program at VIR. Lysenko sketched a number of new investigations, including the effect of high temperatures on the development of cotton plants, and the effect of darkness (photoperiod) on corn and millet. He asked for a number of new assistants and research workers. Both were granted most of what they applied for.[19]

This grant from VASKhNIL was part of a broad support from the central agricultural authorities. The minister of agriculture, Iakovlev, declared in July 1931 that the work of Lysenko had "special importance." Lysenko was therefore to receive support from the ministry for a long list of research projects, including 150,000 rubles to start a new journal.[20]

The first issue of Lysenko's new journal (*Biulleten' Iarovizatsiia*), published in January 1932, contained a report on the vernaliztion of seed grain in 1930, admitting that the vernalization of winter wheat had failed so far. "The fundamental condition for obtaining the necessary effect from spring sowing is the choice of suitable varieties for each region," he concluded (Lysenko 1932a, 57). The task was more difficult than he had imagined. Specific results from 1931 were not reported. Presumably they gave little additional support for the vernalization of either winter or spring wheat.

Despite this lack of positive evidence, the practical trials were expanded in 1932. According to Lysenko (1932b) forty-three thousand hectares were sown with vernalized wheat, mainly spring wheat, and mostly in the Ukraine. A set of three questionnaires was distributed to all participating farms. The first registered the period from sowing until the sprouts emerged from the ground. The second registered when spikes were first formed, and the third the size of the harvest. By August 27 questionnaires representing approximately twenty thousand hectares had been returned. But by far the largest part of this area was represented only by returns on the first questionnaire. There were 772 replies to the first, 77 to the second, and 59 to the third questionnaire.[21] On this material Lysenko built a preliminary report that he published in the August–September issue of his journal (Lysenko 1932b).

Again Lysenko emphasized the speeding up of plant development. He reported that the vernalized plants were on average three to four days earlier in emerging from the ground and in heading, arguing that even such a modest effect could be important. The effect was larger with some varieties than with others. So far only the ordinary local varieties had been used. It was likely that other varieties with a larger effect could be found.

The low percentage of returns on the third questionnaire made Lysenko properly cautious with claims about increased harvest. The large majority of the received reports in this category showed an increase in yield with vernalization, but there were also many cases of lower yields for the vernalized fields. Lysenko used much space to describe successful cases. In his opinion these gave the most correct impression of the possibilities of the method when it was properly applied. One could say that his first report on the 1932 season was quite sober in its verbal form but selective in its choice of facts to present. "The small amount of existing information makes it difficult yet to discuss" the effect on the harvest, Lysenko modestly concluded (1932b, 15).

The need for a closer analysis of the harvest yields was obvious. Increase in yield was, after all, the crucial criterion of the whole trial process. However, a long article by Lysenko filling the October–December issue of the *Biulleten' Iarovizatsiia* treated the speeding up of sprouting and heading in great detail. It also compared the effect on cereals with the effect on a number of other cultivated plants (Lysenko 1932c). The yields were once more ignored.

Later Lysenko did publish some more data on yields in the 1932 trials. He reported, for instance, that comparison of vernalized and nonvernalized spring wheat had been carried out in 240 *kolhozes*. Of these 18 achieved an increase of 3 to 9 *tsentners*[22] per hectare, 38 an increase of approximately 2 *tsentners*, and 127 an increase of up to 1 *tsentner*. In 25 *kolkhozes* the yield was approximately equal, in 30 there was a decrease of up to 1, and in 2 a decrease of up to 2 *tsentners* per hectare. Lysenko maintained that in many cases mistakes in the application of the technique explained the lack of positive results. For instance, vernalized seed were heavier because of the water absorbed. If the same weight of seed was used per area as for nonvernalized this would lead to thinner sowing and too low density of plants (Lysenko, 1935c, 102–104).

But the lack of clear positive results from the extensive trials did not prevent a further expansion of the trial program. Lysenko reported that in 1933 approximately two hundred thousand hectares were sown with vernalized spring wheat. In 296 *kolkhozes* that reported the harvest, the results showed a similar distribution to that of 1932. In 1934 more than five hundred thousand hectares of spring wheat were vernalized and more than 25,000 *kolkhozes* and *sovkhozes* participated. Lysenko calculated the average increase to be 1.2 *tsentners* for 1933 and 1.13 for 1934 for the farms that had reported. This represented a rise in yield of approximately 10 percent (Lysenko 1935c, 105–15). If a similar effect could be repeated for the whole country, it would obviously be an important achievement.

The methodological weakness of Lysenko's investigations should be clear: yields were reported for only a small part of the trials. It is likely that the best results were overrepresented in the reports. Farms with obviously poor results were more reluctant to report. Nevertheless, Lysenko could justifiably argue that the method had not been disproved because there were many cases where the farms had not mastered the technique or had not applied it correctly. In the tumultuous and strained conditions of a Soviet agriculture under collectivization, reliable tests were more difficult to carry out than under normal conditions.

Most of the leading agricultural scientists appear gradually to have accepted, with reservations, the vernalization of seed grain as a useful minor method, useful at least under some conditions. After all, Lysenko had now given up the more spectacular measure of vernalizing winter grain for sowing in the spring, and only claimed some usefulness for the much less radical method of vernalizing spring grain. Wheat growing was the main theme of a VASKhNIL session at Saratov in October 1935. Other methods of increasing the yields received more attention than vernalization did. But there appears to have been little explicit criticism of vernalization and many positive remarks. Tulaikov, for instance, mentioned an interesting result in its favor. He had found that fertilization with nitrogen and phosphorous produced a yield increase of 1.4 *tsentners* without vernalization but three times as much, 4.3 *tsentners*, with vernalization (Tulaikov 1935, 192). But it was also clear to the session that the method had its limitations. According to the VASKhNIL secretary Bondarenko, "the session had carried out an exhaustive assessment of vernalization as an agrotechnical method." On this basis a "precise list would be worked out of the regions where vernalization was the most effective method," as well as "a precise list of the varieties that gave a positive effect with vernalization" (Bondarenko 1935).

A resolution from this session evaluating the results of Lysenko's work on vernalization said that data from four years of trials had demonstrated beyond doubt an average yield increase of more than 1 *tsentner* per hectare, and that the extra work did not exceed half a working day per hectare. The resolution also acknowledged that vernalization had now passed the experimental stage and that responsibility for the continuation of the program had passed on to the agricultural administration.[23] This gave the method an official stamp of approval and made criticism more demanding. The burden of proof had been shifted from those who supported the method to those who opposed it.

The last issue of Lysenko's journal *Bulleten' Iarovizatsiia* came in December 1932. But in 1935 he was able to start a new journal, this time called *Iarovizatsiia* (Vernalization). In the two first annual volumes, 1935 and 1936, the vernalization of spring wheat was still the main topic. Articles gave detailed accounts of individual successful trials, and numerous short reports and letters from workers and leaders of collective farms as well as local administrators and agronomists told encouraging stories about successful trials. Lysenko's journal contained very little critical comparison and analysis of the results. It evaded more specific answers to the crucial questions: *Under what conditions* does vernalization increase yields and *how large* is the increase?

According to Lysenko the yield increase in 1935 was 2.5 million *tsentners*, that is, 250,000 tons (Lysenko and Prezent 1935, 8). An article in the first issue of *Iarovizatsiia* for 1936 shows how this figure was obtained. The total area sown with vernalized wheat in 1935 was estimated to be 2.1 million hectares, and the average increase was estimated to be 1.25 *tsentners* per hectare, or a little above 10 percent. The latter estimate was made on the basis of questionnaires. But as in the earlier trials, the method was quite problematic. For instance, the returned forms covered only 82,000 out of the 2.1 million hectares, and their representativity was not discussed (Rodionov and Filatov 1936, 91).

CRITICISM AND ABANDONMENT OF VERNALIZATION

Eventually more precise tests were carried out by other researchers. In November 1936 VASKhNIL academicians Piotr Konstantinov and Piotr Lisitsyn, together with the Bulgarian geneticist and plant breeder Doncho Kostov, published a critical article about the work of the Odessa institute. The article was mainly concerned with plant breeding and genetics, in particular the new wheat varieties that

Lysenko claimed to have bred in record time. But one and a half pages were devoted to the vernalization of seed grain. After a few introductory words about the "enormous" importance of vernalization and other achievements at Lysenko's institute, the article turned to sharp criticism. It concluded by regretting Lysenko's polemic way of expression. He and his collaborators seemed unwilling to participate in a "rational discussion" aimed to "discover truth" (Konstantinov, Lisitsyn, and Kostov 1936).

The tendency to make a "fetish" of early ripening is harmful, wrote Konstantinov, Lisitsyn, and Kostov. Under some conditions quick development and early ripening is advantageous, and under other conditions it is harmful. The effect varies from region to region. It depends on the variety and the climatic conditions of the specific year. Lysenko and his collaborators were reproached for proceeding in a way that they should have known could produce misleading results. The description of Lysenko's careless methodology amounted to a veiled accusation of scientific misconduct, if not conscious fraud. For instance, in some places the vernalized grain was given better conditions than the nonvernalized, and in other cases negative results were summarily discarded as due to "neglect of the necessary technique for vernalization" (Konstantinov, Lisitsyn, and Kostov 1936, 122–23).

Lysenko, in response, described the criticism as belonging to an obsolete bourgeois science that was about to collapse under the pressure of the new "theory, experience, and practice" that grew from the Stakhanov movement of the state and collective farms. Examples did not need to be elaborated since they could "be read in the press every day, centrally, regionally, and locally." The criticism was also self-contradictory, according to Lysenko, by first approving his work and then completely rejecting it (Lysenko 1936d, 131–32).

With respect to the specific criticism of his claims about increased yields due to the vernalization of seed grain, Lysenko was once again evasive. He referred to earlier publications and made no attempt to present new or better evidence and arguments. For 1936 he could give no full overview of the results because the program had been taken over by the agricultural administration. But he was convinced that the increase in yield was at least ten million *tsentners* (one million tons). To substantiate the claim, he described a few examples from southeast Russia, which he supposed to be the region that the critics had in mind when they wrote about decreased yields due to vernalization (Lysenko 1936d, 132–34). He thus persisted in depending on precisely the type of arguments that Konstantinov et al. had criticized.

Konstantinov did not give up, however. During the winter and spring of 1936 to 1937, he repeated the demand for a more precise testing of vernalization on several occasions. At the big December 1936 conference on genetics and breeding, Konstantinov gave a lecture in which he showed in detail how vernalization under some conditions led to a decrease in yields. During the first part of the talk, Lysenko frequently interrupted, asking for concrete examples of such negative effects. But as Konstantinov continued systematically to present his data, Lysenko grew silent (Konstantinov 1937).

Lisitsyn also used the December 1936 conference to present a detailed criticism. So far we lack precise knowledge of what the results of vernalization are, he maintained, and there is thus no solid basis for Lysenko's claims. He illustrated the situation with a story from ancient Rome: A seafarer had decided to make an offering to the gods before he embarked on a long and perilous voyage. He therefore wanted to find out to which temple it would be most advantageous to bring it. Everywhere he found tablets with the names of those who had made offerings and been saved. But when he asked the priests for the names of those who had made offerings and *not* returned, he got no answer. In Lisitsyn's view, the evidence that Lysenko presented for the effectiveness of vernalization was equally one-sided (Lisitsyn 1937).

In May 1937 Lysenko published a new reply to Konstantinov and Lisitsyn. He held on to his main claims, but gave concessions on some subordinate points. For instance, he accepted that the effect of vernalization could vary with meteorological circumstances and that precise investigations were needed to make the method more effective. But this could not be achieved by Konstantinov's methods with experiments on small plots of 100 square meters, claimed Lysenko. "These questions must be solved in close cooperation with production and the producing kolkhoz peasants" (Lysenko 1937b, 18-19). The results from small-scale experiments, out of contact with agricultural practice, had little weight compared to the positive results that had been achieved on millions of hectares on the collective and state farms, he argued. The total area of Konstantinov's fifty-four experimental plots was only 3.93 hectares! Lysenko pointed out. The fact that vernalization was actually being used on a large scale was in itself an argument for the usefulness of the method (Lysenko 1937b, 17). Perhaps Lysenko was sincerely convinced of the truth of his own claims. It was not just insincere rhetoric when he described the objections as mere hairsplitting and Konstantinov as an obstructionist.

Lysenko also threatened his opponents by alluding to political reprisals. There have been cases where the data from experimental stations have been completely overruled by results from practical farming, he reminded them. Konstantinov ought to consider that in such cases those who stubbornly defended such data were "brushed away from the arena of scientific work" (Lysenko 1937b, 18).

In spite of Lysenko's aggressive counterattacks, Konstantinov and his supporters were not defeated. Lysenko's last sally functioned mainly as a smoke screen for his retreat. One sign of effective resistance was the rather sudden disappearance of the vernalization of seed grain from the pages of *Iarovizatsiia* in 1937. During the following years there were extensive discussions and struggles over the vernalization of seed grain in the agricultural community. This led to simplification and eventually to complete abandonment of the method.[24]

One critical agricultural scientist tells how he met Lysenko at the conference on drought in 1931 and agreed to carry out trials with vernalization at the All-Union Institute of Grain Culture in Saratov. He eventually concluded that the effect of vernalization on spring plants was mainly due to their being sown in a germinated state, and accordingly he suggested that the method be simplified to a mere moistening and germination before sowing, which was a method that had traditionally been used in many places. In 1940 such a method of simply "moistening" the seed was officially included in the government plan.[25] This account corresponds well with the observations of a British Australian plant physiologist who was in Moscow as a diplomat[26] in 1944–1945. He made a special effort to establish the status of the method at that point, but direct questions to Soviet scientific colleagues produced only evasive answers. His clear impression was, however, that what was then practiced as "vernalization" was hardly more than a germination test and that the great campaign for the vernalization of seed grain had been a fiasco.[27]

How extensive was the practice of vernalizing seed grain? Information from Soviet sources was assembled at the Imperial Bureau of Plant Genetics in Cambridge, England, and summed up in the following table:

Year	Hectares
1932	43,000
1933	200,000
1934	600,000
1935	2,100,000
1936	7,000,000
1937	10,000,000

The author was well aware that something happened to vernalization after 1937, and he comments carefully: "More recent data have not been seen; it is not known to what extent, if at all, Russian farmers are now using the method, or whether any centers for treatment and distribution of vernalized seeds have been operating" (Whyte 1948, 9).

The struggle over the vernalization of seed grain shows that it was possible to resist and beat Lysenko when he was forced to argue over concrete and precise practical issues. Through the persistent and detailed criticism from agricultural specialists it became clear to the agricultural administration that Lysenko's method of vernalizing seed grain was grossly overvalued. This process took time, however, and the final outcome was not publicized. Lysenko's defeat was hushed up to the extent that even scientists in neighboring fields continued to speak about the vernalization of seed grain as a useful method. It must at least have been temporarily useful even if abandoned in the end. Lysenko's general reputation for practical contributions to agriculture derived mainly from his vernalization of seed grain. Though the method was abandoned as useless, the reputation continued to promote his career.

Nevertheless, there were good reasons for uncertainty about vernalization as an agricultural technique. It was a complex question. All depended on the reaction of specific varieties to specific climatic conditions, and it took time to investigate the possibility of very favorable results in certain important cases. Vavilov and others therefore had reason to *hope* for positive results from the vernalization of seed grain in the fight against summer drought and winter perishing in southern parts of the country, at least until around 1937. However, there was another part of Soviet agriculture where vernalization seemed even more promising. This was the drive to push agriculture into the Arctic.

AGRICULTURE IN THE FAR NORTH

The economic development of the far north was one of the romantic dreams of Soviet modernization in the 1930s. It was a symbol of man's ability to conquer adverse natural conditions. The support that Sergei Kirov, the party boss in Leningrad, gave to VIR and Vavilov's research projects was to a large extent inspired by this drive to the north. When Kirov was killed in 1934, it was soon rumored that Stalin was behind the assassination getting rid of a troublesome rival. Vavilov wrote a commemorative article in the first bulletin of the reformed VASKhNIL, "Organizer of the Victory of Northern Agriculture," describing Kirov's persistent interest in science including numerous visits to the agricultural research institutes around "Leningrad—the city of science" (Vavilov 1935c).

The special climatic conditions made it much more likely that vernalization techniques could play an important role in the far north than in the traditional agricultural regions further south. The growth period was extremely short, the long days were unsuitable for a proper development of many cultures, and because agriculture was largely new in these regions, there was an acute lack of genetically well-adapted varieties. Furthermore, the exploitation of the natural resources of the north for industrial purposes produced a new market for local production of food. Transportation was difficult and expensive, in particular for vegetables. This was also the type of crop for which vernalization techniques were best suited. The extreme natural and economic conditions of agriculture in the far north thus increased the likelihood that vernalization methods could be economically profitable. Developing methods for cultivation of vegetables was the main task of the Arctic station of VIR in Khibinyi on the Kola Peninsula (Sinskaia 1991, 94).

In November 1931 Vavilov was preparing a lecture on "Problems of Northern Agriculture" for the Academy of Sciences[28] and wrote a long letter to Iogann Eikhfel'd (1893–1989), head of the agricultural experimental station in Khibiny. Vavilov discussed at length various topics of interest for his lecture, among them the importance of testing vernalization methods in the north. One short paragraph in the letter strongly urged Eikhfel'd to follow up Lysenko's work on vernalization: "What T. D. Lysenko has done, and what he is doing, is of quite exceptional interest, and it is necessary that the Polar branch develops this work further."[29]

Eikhfel'd, like Lysenko, was a "pushed up" (*vydvizhenets*). He had been active in the organization of soldiers' and workers' soviets during the 1917 revolution. Afterward he spent four years at the Leningrad Agricultural Institute and then went north to cultivate the Arctic. Vavilov helped him convince the government that this was a project worth supporting. Eikhfel'd was soon fascinated by vernalization and became one of Lysenko's supporters. When Vavilov was arrested in 1940, Eikhfel'd became acting director of VIR.[30]

ENTHUSIASM AMONG BOTANISTS

Lysenko had more success among academic plant physiologists than with shrewd agricultural scientists. Soviet plant physiology at the beginning of the 1930s needed research topics with practical relevance. Vernalization had just the right combination of theoretical and practical interest. It was an important topic on the international research frontier and it was highly relevant to pressing problems of Soviet agriculture. Lysenko's vernalization of seed grain was just one among many possible uses.

The enthusiasm for vernalization as a research topic appears to have reached a high point toward the end of the period of cultural revolution. At a series of congresses and other scientific meetings in 1931–1932, leading botanists vividly expressed this enthusiasm. These public discussions expressed their efforts toward a "socialist reconstruction" of plant physiology. The typical agricultural experts were absent or played a minor role in these discussions.

In September 1931 the Ministry of Agriculture organized a conference to survey possible contributions of science to the increase of agricultural yields. One month earlier the government had issued a decree, "Plant Breeding and Seed Production," which demanded greatly increased efficiency in plant breeding.[31] In Stakhanov fashion, the time used to produce a new variety should be cut from ten to twelve years down to four or five. The possibility of speeding up plant breeding was therefore a main topic at this conference.

Vavilov spoke very favorably of Lysenko's research on vernalization in a lecture called "New Roads in Research on Plant Cultivation." Remarkable new facts on plant development had been discovered in photoperiodism by Garner and Allard and in vernalization by Lysenko. Now an "enormous collective work on the large material of varieties," that is, the World Collection, was needed to derive practical consequences from these discoveries. On the specific question of the vernalization of seed grain Vavilov was more reserved: "There is still no valid basis for entering broad production experiments with a full guarantee."[32]

In his talk Lysenko objected to the common misunderstanding

that his method applied only to the vernalization of winter grain. The effect was much more general, and it was in speeding up the development of spring varieties that the greatest promise lay. He also emphasized that vernalization was in no way opposed to selection. It was not a matter of chosing between the two. This was another misunderstanding that he often had to fight. The two methods could and should work hand in hand, emphasized Lysenko. Like Vavilov, he pointed to the general use of vernalization in controlling plant development.[33]

There was broad agreement that the new physiological discoveries would make plant breeding more efficient. But it was warned, for instance, by Meister, that the demands of the August government decree were impossible to fulfill. To produce a new variety in four to five instead of ten to twelve years was not practically possible. Iakovlev replied tersely: "We cannot wait for ten years."[34]

The Ministry of Agriculture continued its special support of Lysenko at the All-Union Conference on Drought Control in October 1931. Iakovlev ventured to say that Lysenko underestimated the changes his experiments would bring about in agriculture.[35] He declared it a main task for 1932 to find the right varieties of wheat and other cultures for vernalization in the various regions. In 1933, mass trials of Lysenko's method would start "on at least hundreds of thousands of hectares."[36] The congress passed a resolution declaring that Lysenko's research was so important that his works must be published "out of turn," as quickly as possible, and that he should be freed from other duties and given "maximally favorable research conditions."[37] This was while the heady visions of the cultural revolution were still dominating the public media, before the sobering hangover.

Soon after this conference on drought, the Leningrad branch of the Society of Marxist Biologists of the Communist Academy organized a meeting to judge its results. The main speaker was Vavilov. He did not say much about the vernalization of seed grain but developed at length the importance of vernalization in plant breeding. Because most cultivated plants have their origins at southern latitudes, it was often difficult to make them flower under Russian con-

Vernalization | 135

ditions. This was a big obstacle for breeding work. In some cases as much as 90 percent of the potentially useful breeding material was in practice inaccessible for such reasons. To overcome this problem, it was now necessary for the plant physiologists to make deeper studies of photoperiodism and vernalization and to help develop practical methods for the breeders.[38]

In the dicussion, Maksimov reminded the audience that historically the discovery of vernalization "belongs not to Lysenko but to Gassner, the German scientist who published it already in 1918." The merit of "agronomist Lysenko" was that he had applied this "theoretical discovery of Gassner" in a way that could be used by almost any *kolkhoznik*.[39] In his final remarks Vavilov mildly rebuked Maksimov for his condescending attitute toward Lysenko's scientific achievements:

> Let me now turn to the controversial questions (*spornye voprosy*). I believe that N. A. Maksimov somewhat underestimates the work of comrade Lysenko. I believe that they represent a big step. Lysenko not only developed Gassner, he went much further, he took the most different objects, he found that not only lowering of temperature, but also many other factors can be used to speed up development. His approach is very serious, and very new. We agronomists feel that a real revolution (*povorot*) has started. . . .[40]

Lysenko's publications did not give an adequate impression of the width and depth of his approach and results, added Vavilov. Only by "concrete acquaintance on the spot in Odessa can we see how interesting they are."

The impression of the verbatim report from this discussion is that Vavilov went out of his way to support Lysenko against Maksimov's criticism. He identified himself as an "agronomist," like Lysenko. He urged and flattered Maksimov to make him take up the scientific challenges of Lysenko's results:

> I think that it [is] essential that precisely the laboratory of N. A. Maksimov takes on the further investigation of Lysenko's results, for the deeper theoretical analysis, which they need. . . . A broad

physiological approach is needed, which is only in the power of a strong institution like the laboratory led by Professor Maksimov.[41]

A few months later the enthusiasm for Lysenko and his vernalization appears to have reached a high point at a conference on agrophysiology at VIR in February 1932. An extensive report (Anon. 1932) in the scientific journal of VIR presented the conference as a great success for Lysenko. He now received all the attention and praise from the scientific establishment that he had longed for three years earlier, in January 1929. Plant development was the central theme, and Maximov gave the introductory survey lecture. But a second main lecture was given by Lysenko. According to the printed report, he was met with "persisting applause" when stepping to the podium. And during the discussion it was his views and not Maksimov's that attracted the most attention.

Maksimov himself now took a more positive attitude toward Lysenko's achievements. He emphasized that during the ongoing "socialist reconstruction" a fundamental task was to "seek ways of steering the individual development of the plant organism" (Anon. 1932, 16). He mentioned in particular three practical agricultural tasks to be solved by shortening the growth period: extension of important cultures toward the north (into cold areas), extension toward the southeast (into dry areas), and getting two harvests in one year. The manipulation of light regimes was one possible approach. But Maksimov gave most attention to vernalization.

Maksimov stressed the "practical" character of Lysenko's work in contrast to traditional Western science that was separated from practical work. Lysenko's vernalization was a paradigm example of the unity of theory and practice (17–18). He conceded that Lysenko had been basically right in their earlier controversy over the nature of vernalization: "The theory of antagonism has served its century and must be replaced by a new theory" (Anon. 1932, 18).

Lysenko in his lecture used the term "vernalization" to designate a general theory of plant development (Anon. 1932, 20–23). From this generalization it was but a short step further to include plant breeding and genetics.

Dmitrii A. Sabinin (1889–1951) was the first speaker in the debate that followed the lectures of Maksimov and Lysenko. Sabinin was professor and head of plant physiology at Moscow University from 1932. He later became one of Lysenko's severest critics. But on this occasion he was full of praise and revolutionary rhetoric: "Lysenko's work inspires a feeling of edification, this purposeful work, this thoroughly purposeful work, subordinated to the demands and needs of the construction of socialism." He had seen Lysenko's experimental discoveries at both Gandzha and Odessa and considered them to be "an illustrious step forward on the road of our knowledge about the life of plants." Sabinin was particularly impressed by Lysenko's method: He "says that he is no dialectician," but more than anybody else he has demonstrated dialectical materialism through his presentation and "taught us, physiologists, how we must organize our work in order to answer the questions that have brought us here."[42]

V. I. Razumov from VIR's laboratory of plant physiology found Maksimov's "self-criticism" unsatisfactory. Maksimov was still too bound by Western science, for instance, by Gassner, Garner, and Allard. The "necessary critical reworking" was missing. Lysenko's results were "the first real Soviet contribution to plant cultivation." The fundamental mistake of VIR's physiological laboratory was "the incorrect interrelation between theory and practice."[43] A year later, in February 1932, Razumov took over as head of VIR's laboratory of plant physiology when Maksimov was first arrested and then moved to Saratov.

Vavilov had not been present at Lysenko's lecture, but in the discussion he joined the chorus of praise and great expectations. Physiology of development was that part of plant physiology which was making the most progress. "We are in fact approaching the art of transforming the plant according to our will," said Vavilov. Plant breeders are reluctant to make use of the new method of vernalization, but "fortunately Lysenko is close to breeding, is himself a breeder." Vavilov even suggested that the new results would provoke "serious revisions of some fundamental theses in genetics." The new path that Lysenko in particular had helped to open was leading to "great achievements, great practical achievements, great theoretical achievements." Vavilov believed that "this day of our conference will be reflected in all coming conferences. . . . We feel that now we really begin to take possession of plants."[44]

The following speaker was Isaak Prezent, who extracted from Lysenko's work its consequences for general scientific method. The radical self-criticism of Maksimov, admitting that plant physiology had created very little of economic use, was relevant for other branches of biology as well. As examples Prezent mentioned two recent books. One was by the geneticist Iurii Filipchenko; another, by the embryologist M. Zavadovskii. Compared to such people Lysenko had published almost nothing, but if one asked who had produced most "works," there was no doubt, said Prezent, who was then applauded by the audience.[45] The work of Lysenko was radically different from traditional plant physiology. Precisely because "the practical importance of this work was beyond doubt," special attention should now be given to its theoretical implications. To avoid "falling into metaphysics" in the theoretical analysis of the problems of vernalization, it was necessary that Lysenko "received sufficient support from comrades stronger in methodology," argued Prezent (Anon. 1932, 24).

In conclusion, Lysenko remarked with proper modesty that his personal achievements were being overrated. He had not been alone. By himself he could do but little.[46] Only by collective efforts had the work been successfully carried out. Thus Lysenko in his own way echoed the atmosphere of collectivism and Vavilov's call for planning and cooperation both within and between the different research institutions.

The session was not completely lacking in critical comments, however. Toward the end of Lysenko's final remarks there was an exchange with Krasnosel'ska-Maksimova, Nikolai Maksimov's wife and collaborator. She pointed out that Lysenko's method for vernalization of seed grain was considerably more difficult to handle for the peasants than he made it appear.[47]

The day and the session were rounded off by the chairman, Richter, repeating Lysenko's summons to collective future efforts.

Lysenko spoke the truth, if he is to be alone in his further work he will obtain no results. It is necessary, clearly necessary, that we all work together. Wherever we came from, we must be one collective.[48]

The only name mentioned in the eight resolutions from the conference was Lysenko's. Vernalization provided not only the means to combat drought, increase yields, and so on. It also "revolutionizes the attitude to the plant." A research plan was sketched that included the formation of a common brigade of workers from Vavilov's VIR and the Odessa institute to study methods of vernalization (Anon. 1932, 28). In 1932 Lysenko's section of plant physiology in Odessa was heavily involved in the systematic vernalization of material from the World Collection at various geographical locations. This was a program aiming to find "the most suitable varieties for a given region, both for direct agricultural use and for purposes of breeding."[49] Apparently no active brigade with official participation from VIR was formed. But a brigade from the Communist Academy of Leningrad headed by Prezent was in Odessa during the summer of 1932. In January 1933 Lysenko described this event as a great inspiration for his general theorizing about vernalization.[50]

Contact between VIR and Lysenko did develop from the early 1930s. But this did not further the theoretical development and critical scientific evaluation of vernalization that Vavilov had asked for. After Maksimov was arrested in February 1933, the plant physiology section of VIR lacked general theoretical competence. A review of the international response to Lysenko's new techniques of vernalization that Razumov published in Lysenko's journal *Iarovizatsiia* in 1936 was descriptive and scientifically undiscriminating (Razumov and Griuntukh 1936). It lacked critical discussion with respect both to theoretical explanation and the practical usefulness of the techniques.

The conference on agrophysiology in February 1932 marked a new stage in Lysenko's scientific career. He had not only gained the favor of the Ministry of Agriculture and become a public figure, nationally famous as the foremost representative of the new breed of "pushed-up" scientists. He had also received the approval of the scientific establishment and now belonged to the country's most influential experts in the field of applied plant physiology. As late as four months earlier, at the conference on drought control, when Iakovlev had urged him to show less modesty, he had played a quite subordinate role. It is also notable that the first of the two conferences had a concrete practical agenda, while the second one had a more general theoretical, methodological, and science policy nature.

THEORETICAL PROBLEMS OF VERNALIZATION

One year later the frenzy of the cultural revolution was over and sobering afterthoughts mellowed the enthusiasm. On January 9, 1933, a joint meeting was held in Leningrad by the Society of Marxist Biologists, the Society of Natural Historians (*estestvoispytateli*), and the Botanical Society to hear Lysenko speak on "The Theoretical Basis of Vernalization." The chairman of the meeting was Boris Keller. And those in the presidium seated on the podium included Richter, Komarov, Prezent, L'vov,[51] Liubimenko, Maksimov, Sapegin, and Doncho Kostov.[52] The chairman opened by reminding the audience that "[b]otany is not simply the science of plants. It is the science of how to direct plants, how to remake (*peredelivat'*) them for our big goals, to create a new human culture, to create a classless socialist society."[53]

Lysenko gave a very long lecture split by a pause. Toward the end he reported how Prezent's brigade during the previous summer had taught him the dialectical method of experimentation and reasoning. This had clarified his view of genetics. In earlier work he had been unable either to reach clear conclusions on genetic theory or to properly relate it to his own theory of stages. After all, he was a specialist on the stages of development, and the formation of sex cells and their fertilization were included as parts of these stages. The local Odessa geneticists had not been able to clear up his problems. No names were mentioned, but Sapegin was obviously one of them. "They spoke and I could not agree," explained Lysenko. But Prezent had given the magic word: "It was only necessary for Prezent to say 'relative preformation' (*otnositelnaia zadannost'*)," and I understood everything."[54]

At the end Lysenko asked for criticism, "really severe criticism." Criticism of his work was necessary, he explained, because "I consider it to have grown out of a process of criticism," namely, the critical cooperation with collective farmers. "Please criticize without restraint," Lysenko admonished his academic audience.[55]

In the discussion, Liubimenko and Maksimov commented on Lysenko's physiological theories and Sapegin on his genetic ideas.

Liubimenko spoke first. After respectful phrases about Lysenko's "brilliant" discovery and the new scientific discipline that had emerged in its wake, he ventured to criticize Lysenko's sharp distinction between growth and development. He doubted its general validity. Many phenomena did not fit, for instance, the behavior of perennial woody plants, trees, and shrubs, where growth and development could not be separated in the same way. The underlying process taking place in the protoplasm of the individual cells had to be much more complex than Lysenko thought. Liubimenko suggested the existence of a special hormone.[56]

Maksimov's critical remarks were respectful but not without a somewhat condescending undertone. He found that Lysenko's "extremely interesting, and extremely broad lecture" did not contain any finished and comprehensive theory. But "one could see, from his lectures and writings over several years, how a scientific theory was gradually developing, in its own idiosyncratic way, as theories do when they concern the most difficult problems." Maksimov discussed the problem of reversing the effects of vernalization. Lysenko claimed that vernalization was in principle irreversible, while Liubimenko thought that it could be reversed under certain circumstances. Maksimov suggested a mechanism that could make them both appear right, in a sense, but which was clearly beyond the range of Lysenko's thinking and close to Liubimenko's view.[57]

Sapegin was clearly irritated by the length and detailed and repetitive character of Lysenko's lecture. "It is necessary to be very brief," he addressed the audience, "since I, probably no less than you, want to go to bed." His main point was that Lysenko's views were not inconsistent with classical genetics. Sapegin started with an interpretation of the theory of stages invoking the dialectical law by which the accumulation of quantitative differences leads to a qualitative jump. In his view Lysenko's results were a welcome blow to traditional "formal genetics," namely, the kind of genetics that establishes the number of genes by "purely external assessment of the organism." What Sapegin rejected was a simple one gene–one character relationship. In his opinion Lysenko's results demonstrated the fundamental difference of the genes from the external properties of the organism (the phenotype) and the complexity of the genetic effects on the plant. He emphasized that the phenotypic effects of the genes changed in the course of development of the individual organism and differed with varying environmental circumstances.[58] In other words, Sapegin professed a sophisticated standard classical genetics, as it had been developed, for instance, by Johannsen and the Morgan school. But Lysenko was more comfortable with Prezent's simple formula.

This meeting in January 1933 shows how leading plant physiologists were critically distancing themselves from Lysenko's theory of stages. They continued to praise his experimental discoveries in the field of vernalization, and his theory of stages was treated with respect as an important step forward. This theory had helped form a general perspective on the problems of plant development that was formerly lacking. But it was an insufficient basis for further research. What was now needed were biochemical theories involving hormones and other chemically defined substances.[59] The meeting also shows Lysenko groping his way toward a theory of genetics. Its content was still flexible, and some geneticists, such as Sapegin, were trying to convince him that his ideas could best be clarified within the framework of classical genetics.

THE INTERNATIONAL RESPONSE

In the early 1930s there was an international wave of interest in vernalization. This wave was stimulated by Lysenko's work, which recieved much international attention. Numerous reports appeared in agricultural and biological journals in many countries. Among the first presentations of Lysenko's ideas outside the Soviet Union was an article by Sapegin (1932) in *Der Züchter*, a German journal for plant and animal breeding. Sapegin mentioned both the potential importance for plant breeding and the possibility of a practical agricultural technique. *Der Züchter* published a second article on the vernalization of cereals in early 1933, reporting the main results from the 1932 volume of Lysenko's *Bulleten' Iarovizatsiia* without any critical analysis. On the crucial question of increase in yields, the article simply said results were promising and confirmed expectations from preliminary experiments (Nerling 1933).[60]

But there were also more critical reactions. Two American scientists (McKinney and Sando 1933a) published a short note about Lysenko's work on vernalization together with a paper presenting their own results from experiments started in 1928–1929. Like Maksimov, they disagreed with "the view of Lysenko that temperature is the primary controlling environmental factor in earliness under field conditions." They also pointed out that methods quite similar to Lysenko's had been used in American agriculture in the mid-nineteenth century. Their experimental results indicated that the vernalization of winter wheat was unlikely to become a useful practical measure in the United States (McKinney and Sando 1933b, 172–73). This conclusion was confirmed and extended in experiments published in September 1934 by the United States Department of Agriculture: so far spring wheat had proved better than vernalized winter wheat, and vernalized spring wheat no better than nonvernalized (McKinney et al. 1934, 6).

However, the most influential Western source of knowledge about Lysenko's work on vernalization was a British institution, the Imperial Agricultural Bureaus. This was a scientific institution but not doing its own research. Its purpose was to collect and distribute information about recent research results from all over the world with a special concern for the needs of the British empire. Two branches of this institution covered plant breeding. At the Imperial Bureau of Plant Genetics, Herbage Plants, in Aberystwyth, Wales, the work was directed by Dr. R. O. Whyte. In Cambridge there was a branch for crops other than herbage plants, with Vavilov's former host, Prof. R. H. Biffen, as director and Dr. P. S. Hudson as vice director and executive officer. Hudson and Whyte saw it as a main task to inform about the rapidly expanding Soviet agricultural research both because of its many interesting results and because the Russian language represented a serious barrier to most Western scientists.[61] These two Imperial Bureaus of Plant Breeding were established in 1929.

In March 1933 the two bureaus jointly published a special bulletin, *Vernalization or Lysenko's Method for Pre-Treatment of Seed* (Whyte and Hudson, 1933). This bulletin of approximately thirty pages described Lysenko's theory of stages in the development of plants as well as his experimental methods. Some of his main results from experiments with temperature and photoperiodic effects on various plants—wheat, potatoes, soybeans, and so on—were given. The distinction between growth and development was emphasized as the central point of Lysenko's theory. Growth was a quantitative increase in weight, development a qualitative change in the nature of the plant, that is, its transformation from one stage to another. The bulletin was a neutral report. It contained no evaluation or discussion of Lysenko's claims. According to Whyte this was a conscious line of the bureaus. As institutions without their own researchers, they lacked the expertise to conduct a thorough critical evaluation. Their task was only to give detailed, sober, and reliable information about Soviet agricultural science. It was then up to the reader to judge its value.[62]

The 1933 bulletin on vernalization was in great demand and represented a big success for the young Imperial Bureaus of Plant Breeding. Their efforts in translating and abstracting the Soviet literature on vernalization was intensified. A new comprehensive survey of the research on vernalization in both the Soviet Union and other countries, *Vernalization and Phasic Development in Plants*, was published in December 1935. The first bulletin, from 1933, had stimulated a worldwide wave of research on vernalization, and this new bulletin described the results that were pouring in.

The central theme of the 1935 bulletin was Lysenko's theory of stages. The hormone theory of Liubimenko was presented as the main alternative—an alternative that could apparently account better for the reversibility of vernalization processes and for the ability of special light regimes to substitute for the cold requirement. The bulk of the 150-page bulletin, however, consisted of short summaries of a large number of experimental papers, most of which were Russian. Again, the exposition was highly descriptive. Presumably, some critical standards were used in selecting from the available Russian material. But there was no explicit attempt to evaluate the work that was summarized. Since very few of the readers could check the original publications for details, this cautious policy of pure description lent to Lysenko's results and speculations a misleading plausibility.

The initiator and driving force of the 1935 bulletin was Whyte. But half a year before it was finished, the central leader of the Imperial Bureaus, Sir David Chadwick, enlisted Frederick Gregory as a special consultant.[65] Gregory had worked on vernalization problems from 1931 onward and was by 1935 perhaps the foremost scientific expert on this topic in Britain. In 1937 he became professor of plant physiology at Imperial College, London.[64] Two months before the bulletin was finished, Whyte offered Gregory coauthorship.[65] But eventually the bulletin was published without any personal authorship. Gregory's assistance was acknowledged in Chadwick's foreword, and on reception of the finished work he wrote to Whyte that he felt "quite proud to have been associated with its production."[66] Gregory probably did not want to be too closely associated with the work of Whyte and Hudson, whom he felt to be somewhat uncritical in their attitude toward Lysenko and not quite on the same scientific level as himself. This is indicated in a letter that Gregory wrote to J. B. S. Haldane some years later warning him not to take Lysenko's plant physiology too seriously.[67]

Both Hudson and Gregory spoke on vernalization at the Sixth International Botanical Congress in Amsterdam in September 1935. The session was called "Influencing the Cycle of Development in Plants" and was a combined session of the plant physiology and agronomy sections of the congress. Hudson lectured on

"Vernalization, Lyssenko's Yarovization of Seed." He described Lysenko's experimental technique and his theory of stages, focusing on the separation of growth and development that allowed a "pretreatment of seed." Other investigators had found that low temperature is not obligatory for the flowering of winter cereals, but can be substituted by a short day treatment, Hudson noted. Gregory's lecture, titled "Photoperiodicity Problems in Cereals," described his own investigations of developmental processes at the growing point and discussed possible causal mechanisms of vernalization. He mentioned that Maksimov had shown that the "cold requirement" was not necessary, that his own experiments showed no "causal effect of the carbon-nitrogen ratio," and that the main difficulty of the hormone hypothesis was to explain the predetermination effect of temperature.[68] The two papers illustrate the difference between a purely reporting author and one with an active involvement in scientific research.

In the discussion after these lectures one of the organizers of the congress told how "the news about the Russian *jarowisation* work about a year and a half ago caused rather a stir in Netherlands agricultural circles, as in neighbouring countries." He thanked Gregory and Hudson for their "excellent contributions." Both Gassner and Richter had also "promised a paper for this session, but were prevented from coming."[69]

Maksimov had admitted his own mistakes and given high praise to Lysenko's experimental discoveries and practical achievements. But he also continued as a persistent critic of Lysenko's general theories of plant development. In December 1934 he published a review article on vernalization as a bulletin of the Imperial Bureaus of Plant Breeding. Here Maksimov once again acknowledged the great significance of Lysenko's method of accelerating plant development and respectfully described his theory of stages. But he also presented alternative theories and objections by other Soviet scientists. His concluded that some of them would probably prove invalid, but others "will certainly prove justified" and necessitate "a number of changes and corrections in Lysenko's theory" (Maksimov 1934, 13). Among other things, Maksimov mentioned the hormone theory, which he attributed to Liubimenko, and results from several researchers showing that the need for a cold period was not as rigorous as Lysenko maintained. Under certain special conditions, it could even be completely dispensed with, as shown, for instance, by the Soviet plant physiologist Chailakhian. Though Maksimov's formulations were circumspect, his view was clearly that either Lysenko's qualitative theory of stages had to be transformed into a quantitative chemical theory or it would have to be given up. Its present form was too crude to be of real scientific interest.

In a Russian review of foreign literature, Chailakhian described the results of McKinney and Sando and their conclusions that temperature is not the fundamental control factor that Lysenko took it to be. His own work pointed toward the same conclusion, wrote Chailakhian. And he maintained that Lysenko's formula for the temperature dependence of the length of developmental phases would have to also include the length of light periods to be applicable to a broad range of latitudes (Chailakhian 1934).

The numerous reviews of the 1935 vernalization bulletin from the Imperial Bureaus is an interesting gauge of Lysenko's status among Western scientists at this point. The comments of M. A. H. Tincker, British plant scientist, are representative. He praised the bureaus for having attracted Western attention to Lysenko's "important theory" and having made it so readily accessible. He considered Arctic agriculture to be the main area of economic application, and he emphasized the scientific interest and possible future economic importance of Lysenko's work. But one can perhaps discern a reservation with respect to the real practical achievements so far: "The further possibilities claimed are extremely interesting: it remains to be seen whether practice will uphold the theory."[70]

An anonymous review in the British science journal *Nature* was positive both in its evaluation of the service done by the Imperial Bureaus and in its assessment of Lysenko's practical achievements: "In some countries and with certain crops, notably cereals, practical success seems evident."[71] In the American journal *Ecology* the attitude toward Lysenko's work was also positive, emphasizing the importance of further research: "The various investigations give results which support Lysenko's theories to a varying extent but sufficiently to prove that vernalization is worthy of the careful attention of all plant scientists."[72] More critical both of Lysenko and of the Imperial Bureaus was the review in the German journal *Angewandte Botanik*. The reviewer complained about the "the present undoubtedly overrated importance of 'jarovisation'" and the uncritical selection of literature for the bulletin.[73]

The mediation work of the Imperial Bureaus stimulated further experimental testing. The Plant Breeding Institute in Cambridge concluded that vernalization produced considerable change in the development of some varieties of wheat, barley, and oats (Bell 1936), but the "[g]ross yield per plot was unaffected" (Bell 1937, 392). Since Cambridge does not usually have drought problems, this result was not necessarily contrary to Lysenko's claims. But it nevertheless served to subdue the first enthusiasm. Hudson in 1936 accepted the usefulness of the vernalization of seed grain in certain parts of the Soviet Union, but held that its applicability to other countries "remains a matter of much doubt." One possible profitable application under English circumstances was to "vegetables and various market garden crops, where a difference of a day or two in date of maturity" may be economically important (Hudson 1936, 539–41).

SUMMING UP

I have shown in this chapter how vernalization in the early 1930s was a hot topic in agricultural research, not only in the Soviet Union but worldwide. The research covered a broad spectrum from the highly theoretical to the immediately practical. In the Soviet Union emphasis was on practical application, but there was also a strong belief in the power of general theory as the most effective guide for practical work. At the practical end were the agricultural authorities. At the theoretical end were the plant physiologists. The latter recognized that Lysenko had made some significant additions to scientific knowledge. His experimental discoveries were important even if his theories were crude and outdated. In purely scientific discussions Lysenko's ideas usually played little role. It was in general historical surveys and other more official policy-colored contexts that his work received attention from leading scientists. However, this much at least is clear: vernalization had become an important research topic in international plant physiology and Lysenko had been instrumental in setting it on the agenda.

Lysenko's scientific reputation and status as a plant physiologist was consolidated in 1932–1935 by ample recognition abroad as well as within the Soviet Union. At home the image of Lysenko as the hero of a modern, nonacademic, and practically oriented kind of science fitted particularly well with the Soviet doctrines of a new proletarian science. But no doubt this image met sympathy in the West as well.

The enthusiasm of Russian plant scientists for Lysenko's work appears to have reached a high point in 1932, toward the end of the period of cultural revolution. At this time the ideas about a radical renewal of scientific methodology inspired by dialectical materialism and the unity of theory and practice were still fresh and appealing. The widely applied vernalization of seed grain to counteract summer drought appeared attractive and convincing to many scientists into the middle of the 1930s. For instance, Vavilov accepted that the vernalization of spring grain in 1932–1933 had increased yields (Vavilov 1934a, 15), and Sapegin reported an average increase of 10 to 15 percent owing to vernalization (Sapegin 1935, 812). But, as we have seen, in 1936–1937 the method came under heavy expert criticism, and it was quietly abandoned toward the end of the 1930s.

However, there were also other interesting possible applications of vernalization. Among these, expanding agriculture to the north and making plant breeding more efficient appeared particularly promising. Vavilov saw vernalization as a crucial instrument in producing the new plant varieties that were so sorely needed in Soviet agriculture. This application created a practical link between vernalization and genetics in addition to the theoretical connection that Lysenko saw.

NOTES

1. The Russian term *iarovizatsiia* first emerged in the middle of 1929 in connection with the experiment of his father, claimed Lysenko (1932c, 3). The English translation "vernalization," derived from the latin term *verna* for spring, was introduced by P. S. Hudson of the Imperial Bureau of Plant Genetics in Cambridge and used for the first time in a brochure that described the main features of Lysenko's theory and experiments (Whyte and Hudson 1933).

2. See, for instance, the standard textbook by F. B. Salisbury and C. W. Ross, *Plant Physiology*, 3rd ed. (Belmont, CA: Wadsworth, 1985), pp. 412–16. Here vernalization is defined as "low-temperature promotion of flowering."

3. "Otkrytie agronoma Lysenko" (The discovery of agronomist Lysenko), *Pravda*, July 21, 1929, p. 4. This early report tells that the seed grain was treated by Lysenko junior in Gandzha and then "sent 3,000 km to the North" to be sown by Lysenko senior. According to later statements by representatives of the Ukrainian Ministry of Agriculture, for example, in *Ekonomicheskaia Zhizn'* on August 4 and in *Pravda* on October 8, the seed grain was prepared by Lysenko senior according to the theories of his son but more or less on his own initiative. Lysenko senior had moistened seed grain in his own hut and kept them in a snow drift for two months before sowing.

4. A. Shlikhter, "O poseve ozimykh kultur vesnoi" (On the sowing of winter plants in the spring), *Pravda*, October 8, 1929, p. 2.

5. "Otkrytie agronoma Lysenko," p. 4.

6. A. Shlikhter, "O poseve ozimykh kultur vesnoi. (Otkrytiia agronoma T. D. Lysenko)." *Pravda*, October 8, 1929. p. 2.

7. Minutes from a session of the scientific collegium (*nauchnaia kollegiia*) of the Institute of Applied Botany together with participants of a meeting on testing of winter hardiness of winter cultures. TsGANTD–St. Petersburg, f. 318, o. 1, d. 230.

8. See Levina (1991, 234).

9. TsGANTD–St. Petersburg, f. 318, o. 1, d. 230, l.96ob–97ob.

10. L. Emelianenko, "Iarovizatsiia ozimykh kul'tur," *SG*, October 10, 1929, described Lysenko's theory and some main experimental results, the speeding up of development in spring grains, and the quantitative determination of cold requirements in winter grains. This cold requirement was explained according to Lysenko's formula $n = A/B - t$ (explained in chapter

3). The lower the temperature (providing it was above 0) the less time would the vernalization take. This is also the first use of the term *iarovizatsiia* that I have found in print.

11. "Iarovizatsiia ozimykh," *SG*, November 13, 1929, p. 3.

12. "Vesennii posev ozimykh otkryvaet shirokie perspektivy" (Spring sowing of winter varieties opens broad perspectives), *SG*, November 19, 1929, p. 3.

13. *SG*, November 19, 1929, p. 3.

14. T. D. Lysenko, "V chiom sushchnost' gipotezy 'ozimosti' rastenii?" (What is the essence of "winterness" of plants), *SG*, December 7, 1929, p. 3.

15. "Otvet kritikam," *SG*, December 19, 1929, signed "Narkomzem Ukrainy" (Ukrainian minister of agriculture).

16. T. D. Lysenko, "Iarovizatsiia v khoziastvennykh usloviiakh" (Vernalization in conditions of production), *SZ*, July 2, 1930, p. 5.

17. "Posevi ozimykh vesnoi po metodu agronoma Lysenko sebia opravdali," *SZ*, February 24, 1931, p. 3. See also the later report printed in Lysenko's new journal, *Biulleten' Jarovizatsiia* (Lysenko 1932a).

18. TsGAE, f. 8390, op.1, e.90, l.1–3. Protocol of "Meeting of the Bureau of the Presidium," February 20, 1931.

19. TsGAE, f. 8390, op. 1, e. 116, l. 12–15, 25–26. Maksimov, on behalf of VIR, applied for 10,400 rubles, which would also cover investigations of "biotonization" of seed. Lysenko, together with the director of the Odessa institute, Stepanenko, applied for 43,225 rubles.

20. See *BIa* 1 (1932): 71–72.

21. Lysenko does not say how many questionnaires were distributed or how many potential respondents there were.

22. 1 *tsentner* is 100 kg.

23. *BV* 11 (1935): 7.

24. Conversation with the plant physiologist Adolf T. Makronosov, corresponding member of the Soviet Academy of Sciences, in Moscow December 5, 1986. Makronosov experienced the struggle over vernalization as a boy at a Siberian collective farm where his father was an agronomist. The abandonment of vernalization toward the end of the 1930s is confirmed by Medvedev (1969, 15–16), who claims that "vernalization of spring cereals fell off sharply even before the war."

25. Autobiographical manuscript by I. M. Vasil'ev, consulted at the Timiriazev Institute for Plant Physiology in Moscow, December 5, 1986.

26. Eric Ashby was scientific attaché at the Australian Legation.

27. Conversation with Ashby in Cambridge, England, May 20, 1983.

One of his sources for this judgment was an interview with the plant breeder N. V. Tsitsin, who earlier had cooperated with Lysenko. See also Ashby (1947), for instance, p. 115.

28. Letter from N. I. Vavilov to I. G. Eikhfel'd, October 19, 1931, VL 2, p. 132.

29. Vavilov to Eikhfel'd, November 11, 1931, VL 2, p. 133.

30. Eikhfel'd has described the first ten years of his agricultural experimental work on the Kola peninsula in a book with a preface by Vavilov, *Bor'ba za krainii sever. Kratkie itogi raboty poliarnogo otdeleniia Vsesoiuznogo Instituta Rastenievodstva, 1923–1933* (Leningrad, 1933). He participated in the reorganization of the Estonian Academy of Sciences after the Second World War. A short biography was published for his ninetieth birthday in 1983, "Nauchnaia, nauchno-organizatsionnaia i obshchestvennaia deiatel'nost' Iogana Gansovich Eichfel'da," *Izvestiia Akademii Nauk Estonskoi SSR* (Biology) 32, no. 1 (1983): 3–8. A more romantic book-length biography is A. Lozhechko, *Pioner severnogo zemledeliia* (Moscow, 1969), which tells the story about the building of large new *sovkhozes* on the Kola from 1931 onward.

31. "O selektsii i semenovodstve," *Pravda*, August 3, 1931.

32. N. I. Vavilov, "Novye puti issledovatelskoi raboty po rastenievodstvu," *SZ*, September 13, 1931.

33. "Govorit agronom Lysenko," *SZ*, September 13, 1931.

34. "My vziali pervuiu krepost'...," *SZ*, September 13, 1931.

35. "Dnevnik konferentsii," *SZ*, October 29, 1931, p. 1.

36. "Selektsiia—vazhneishie oruzhie v borbe s zasukhoi," *SZ*, October 30, 1931, p. 1.

37. See *BIa* 1 (1932): 80.

38. ARAN, f. 1593, o. 1, d. 190, l.11ob–12. Folder 1593 contains the papers of I. I. Prezent, who was at the time president of the Society of Marxist Biologists of the Communist Academy.

39. ARAN, f. 1593, o. 1, d. 190, l.28.

40. ARAN, f. 1593, o. 1, d. 190, l.52.

41. ARAN, f. 1593, o. 1, d. 190, l.52–52ob.

42. ARAN, f. 1593, o. 1, d. 192, l.1–3.

43. ARAN, f. 1593, o. 1, d. 192, l.6–6ob.

44. ARAN, f. 1593, o. 1, d. 192, l.27–29.

45. ARAN, f.1593, o. 1, d. 192, l.29ob–30.

46. ARAN, f. 1593, o. 1, d. 192, l.34ob.

47. ARAN, f.1593, o. 1, d. 192, l.36ob.

48. ARAN, f. 1593, o. 1, d. 192, l.37.

49. "Otchiot o rabote Ukrainskogo instituta selektsii za 1932 god," TsGANTD-St. Petersburg, f. 318, o. 1, d. 495, l.23.

50. ARAN, f. 1593, o. 1, d. 193, l.32-33.

51. Sergei Dmitrevich L'vov (1879-1959) professor and head of the kafedra of plant physiology at the University of Leningrad 1931-1959. Corresponding member of the Academy of Sciences from 1946.

52. Doncho S. Kostov (1897-1949) was a Bulgarian geneticist who worked at the Institute of Genetics of the Academy of Sciences from 1932 to 1936 and later as professor at Leningrad University from 1939 to 1947.

53. ARAN, f. 1593, o. 1, d. 193, l.2.

54. ARAN, f. 1593, op. 1, d. 193, l.32-35.

55. ARAN, f. 1593, o. 1, d. 193, l.34.

56. ARAN, f. 1593, o. 1, d. 193, l.35-38ob.

57. ARAN, f. 1593, o. 1, d. 193, l.42-43obd

58. ARAN, f. 1593, o. 1, d. 193, l.40-40ob.

59. See, for instance, the historical account and assessment of Skripchinskii (1969).

60. For a survey of the early reactions to Lysenko's work on vernalization in the international scientific literature, see Razumov and Griuntukh (1936).

61. On the history of the Imperial Bureaus of Plant Genetics, now the Commonwealth Bureaus of Plant Breeding and Genetics, see W. R. Black, "Imperial Agricultural Bureaus," *Journal of the Ministry of Agriculture* (August 1929): 461-67; P. S. Hudson, "The Imperial Bureau of Plant Genetics," *Journal of Scientific Agriculture* 16 (September 1935-August 1936): 549-52; R. H. Richens, "The Commonwealth Bureau of Plant Breeding and Genetics," University of Cambridge School of Agriculture Memoirs, no. 41, Cambridge 1969.

62. Interview with R. O. Whyte in Paris June 18, 1983.

63. Letter from R. O. Whyte to F. G. Gregory, June 20, 1935. This letter and other correspondence involving R. O. Whyte and P. S. Hudson was studied in the archive of the Commonwealth Bureau of Plant Breeding and Genetics, Cambridge (former Imperial Bureaus of Plant Breeding) in 1982-1983.

64. "Frederick Gugenheim Gregory, 1893-1961," *Biographical Memoirs of Fellows of the Royal Society* 9 (1963): 131-53.

65. Letter from R. O. Whyte to F. G. Gregory, September 23, 1935.

66. Letter from F. G. Gregory to R. O. Whyte, January 23, 1936.

67. Letter from F. G. Gregory to J. B. S. Haldane, August 21, 1947. Haldane Papers, D. M. S. Walton Library, University College, London. Gregory writes that "no one in this country except Dr. Purvis and I have done any research on vernalization, and I can assure you that you can discount a great deal of the loose talking and writing which appears from time to time in book form and as articles by people whose sole claim to knowledge on this matter is derived from publications, often in a language which they do not understand. Perhaps you may guess whom I am referring to, but I do not particularly want to be prosecuted for libel, so I shall mention no names." It seems likely that Hudson and Whyte are included among those Gregory does not want to mention by name.

68. *Zesde Internationaal Botanisch Congres, Amsterdam, 2–7 September, 1935, Proceedings*, vol. 2 (Leiden, 1935), pp. 18–21.

69. *Zesde Internationaal Botanisch Congress*, vol. 1, pp.157–58. O. de Vries was vice president of both the executive committee and the organizing committee of the congress. The Soviet plant physiologist N. G. Kholodny also participated in the congress discussing the role of hormones in the germination of seed in a session on plant hormones (vol. 2, pp. 271–72).

70. Review of *Vernalization and Phasic Development in Plants*, in *Journal of the Royal Horticultural Society* 61 (1936): 185–86.

71. "Practical Applications of Vernalization," *Nature* 137 (March 7, 1936): 408–409.

72. G. D. Fuller, "Vernalization," *Ecology* 17 (1936): 298–99.

73. *Angewandte Botanik* 16 (1936): 502–503. Review by "Voss, Berlin-Dahlem."

Chapter 6.

From Problems of Plant Breeding to Controversy in Genetics

Plant breeding was Lysenko's bridge from physiology to genetics. During the early 1930s vernalization techniques and the theory of stages were main tools in the campaign to improve Soviet plant breeding. Nikolai Vavilov became enthusiastic about the possibilities that Lysenko's work opened for the practical exploitation of the World Collection. In the application of vernalization to the World Collection the two had a strong common interest.

VAVILOV'S SCIENCE OF PLANT BREEDING

A manual of plant breeding, *Theoretical Foundations of Plant Selection*,[1] published in 1935 presented the state of the art as perceived by Vavilov and his collaborators. It presented the principles of international plant breeding as based on classical Mendelian genetics, including current scientific doubts and uncertainties. Particularly relevant for our purposes are the book's views on the nature of plant breeding ("selection") as a science, its ideas about biological inheritance, and the place that it gives to Lysenko's work on vernalization.

In the first chapter, "Selection as a Science," Vavilov defines the subject: "Selection, as a science, is the extraction of varieties according to the needs of man" (*Teoreticheskie osnovy* 1935, 6). Its purpose is to help man in shaping nature according to his needs: "selection is evolution directed by the will of man" (7). Selection is distinguished from the "theoretical science" of genetics that aims simply to describe and understand the phenomena of biological inheritance. As a theoretical science genetics can chose to work with the organisms that it finds most convenient, for instance fruit flies (6).

Vavilov had reasons not to use the terms "pure" and "applied" in his characterization of genetics and plant breeding. The term "pure science" had negative connotations of a bourgeois view of science in the Soviet debate. Still, these terms, as they were used internationally at the time, appear well suited to describe his way of thinking.[2] For Vavilov, selection was an applied science in the sense that it should serve political aims. Its success or failure was ultimately dependent on how well these aims were served. But even an applied science had an important measure of autonomy owing to its basis in pure (theoretical) science. Knowledge about nature is the best foundation for successful practical activities. Agriculture or other economic activities that neglect relevant theoretical knowledge will be less successful than they could have been. For Vavilov, progress in basic biological knowledge was a main factor in the progress of applied science: we are interested in theoretical work, "problems of the gene, theory of mutation, theory of hybridization, problems of phenogenetics, because we do not doubt that strengthened theoretical work will give new stimulus to selection." He also saw a socialist type of society as particularly favorable for a "unity of theory and practice" that would promote "formation of strong theory" (14). This was part of his ideology of planning, in society at large as well as in science.

Selection is still very much an art, wrote Vavilov; it is only in the first stages of becoming a science (4). Its basis is Darwin's theory of evolution by natural selection as elaborated with the help of modern genetics. Wilhelm Johannsen's theory of pure lines and his analysis of the relationship between phenotype and genotype was a basic contribution. Earlier practical selection did not consider the fundamental difference between the properties of an organism and its genes. Modern genetics has taught us how changes in the environment can radically change the properties of an organism. Selection must take account of the possibilities of big changes in the environment, for instance, through chemical fertilizers, irrigation, or vernalization. "Vernalization can radically alter the phenotype even converting perennial plants into annual, late forms into early, with the corresponding changes in all their external properties" (10). Vavilov was critical of tendencies in classical genetics to forget the second term in Johannsen's equation: genotype + environment = organism. Like Sapegin, he thought that vernalization demonstrated particularly clearly the fundamental importance of the environment for our general understanding of genetics.

A particularly promising method of plant breeding, intimately related to the discoveries of theoretical genetics, was hybridization. There was, however, one difficult and crucial problem, according to Vavilov, namely, the choice of parent pairs for the hybridization procedure. So far the choice had been made more or less intuitively by the breeders. Extensive description and analysis of each plant species was needed to provide the breeder with a more sound knowledge basis for the choice of pairs. The handbook called this "special," as opposed to "general," genetics. To become really useful for selection, genetic research had to become much more involved with special genetics, argued Vavilov (11). The second and third volumes of the treatise were concerned with questions of "special selection" (*chastnaia selektsiia*).

A short period of growth from germination to maturity was the key to early ripening and thus to meeting summer drought as well as the short summers of the north. A chapter titled "The Problem of the Vegetation Period in Selection" (Basova et al. 1935) discussed the choice of pairs with respect to the length of the growth period. Until recently, research on the genetic determination of the time to maturation had focused on single genetic factors, neglecting the complex interaction between factors and their relations to internal as well as external environment. Only Lysenko's theory of stages provided the correct approach "for solution of the problems of the vegetation period," argued this chapter. As its authority it quoted

Vavilov, who had stressed the importance of Lysenko's "comparatively simple method" in analyzing the World Collection and making it useful. The purpose of this chapter was to show how a solution to the problem of the vegetation period might be reached "on the basis of Lysenko's work on vernalization" (864–65).

Extensive studies of the World Collection were carried out at the VIR branch in Detskoe Selo using vernalization as a main method. One important question was to what extent yield and vegetation period were properties with an independent genetic basis. If this was the case, it meant that by means of hybridization the high yield of a late-ripening variety could be combined with the early ripening of a variety with low yield (*Teoreticheskie osnovy* 1935, 881–88). Lysenko's method for the choice of pairs, based on his theory of stages, was considered to be useful and quite compatible with the principles of classical genetics. However, variation in the progeny from hybridizations appeared to be considerably larger and more complex than he claimed (889–90).

Andrei Sapegin, in another paper, elaborated on the use of vernalization in plant breeding. He envisaged three main uses: in the choice of breeding material, in the choice of pairs for hybridization, and in speeding up the selection work. For instance, Sapegin argued that in order to know the genetic potential of a plant, we need to know which properties it will develop under different conditions. It is quite possible that such a broad investigation will give us precisely those properties that we need in our breeding program. Vernalization is a necessary tool in such work (813–14). It can also be very helpful in speeding up the breeding work, either by growing many generations in quick succession or by increasing the amount of seed from each plant, according to Sapegin (815–16).

Sapegin's plea for the value of Lysenko's work on vernalization is remarkable for a couple of reasons. First, his clear interpretation of Lysenko's theory of stages makes it appear as centrally important to plant science. Second, he once more emphasizes the implications for general genetics. Vernalization demonstrates the falsity of a simplistic view of the relationship between genes and phenotypic properties, a view that was common in classical genetics:

The present simple projection from the mature plant directly to the genotype, passing over the whole development of the organism, must in the majority of cases lead to false conclusions (816).

Similar criticism became a favorite weapon of Lysenkoists in their struggles with geneticists. Sapegin's intention was to interpret Lysenko's results according to a sophisticated version of classical genetics without compromising on its basic principles. Like Vavilov, he saw the complexities of a practical science like plant breeding and warned against direct application of simple theoretical models. Lysenko's experience held important lessons for the plant geneticist and breeder even if his ideas about heredity were unacceptable.

Theoretical Foundations of Plant Selection gave a prominent place to Lysenko's achievements. A reader without prejudice against Lysenko can hardly avoid the conclusion that he was the single scientist who had contributed most to the development of new methods in plant selection during the preceding five to ten years, not only in the Soviet Union but in the world. This was probably also the sincere opinion of Vavilov, Sapegin, and most of the other authors. But it is important to note that only his physiological techniques and theories were so highly respected. There were numerous reservations against Lysenko's genetic ideas.

VERNALIZATION AS A TOOL FOR PLANT BREEDING

In the early 1930s there was broad agreement among researchers and agricultural officials that the best way to overcome the problems of winter killing and summer drought was to breed new and more resistant grain varieties. Vernalization of seed grain could only be a temporary measure until such improved varieties were available. The government decree of August 1931, demanding that new varieties now be made in four to five years instead of the former ten to twelve years,[3] expressed the urgency of the agricultural situation. Even if most plant breeders found the decree quite unrealistic, there was broad agreement that considerable gains could be made by developing new techniques that made it possible to grow the plants continuously year-round and independently of natural seasons.

On March 29, 1932, a little more than one month after Lysenko's triumphs at the agrophysiology conference in Leningrad, Vavilov sent two letters to the Ukrainian Institute for Selection and Genetics in Odessa. One letter was for the director of the institute and the other was for Lysenko. Both letters were quite formal, indicating that Vavilov had so far had little personal contact with any of them. Vavilov explained that VASKhNIL had been asked by the minister of agriculture, Iakovlev, to assist Lysenko with "all possible cooperation" in his research on vernalization, and that the presidium of VASKhNIL had chosen Vavilov himself to take charge of the assignment. From both the director and Lysenko he asked for information about mass trials and laboratory research on vernalization, and what equipment was needed to make the work more efficient. To Lysenko he also wrote that an "International Congress of Genetics and Selection" would take place in the United States in August and that Iakovlev had said the Ministry of Agriculture would support his participation to give a lecture and present a poster.[4] Apparently Lysenko made no response to this invitation, and he did not go to the congress.

Taken alone, this letter to Lysenko can be interpreted simply as Vavilov carrying out VASKhNIL decisions, which had been taken under political pressure from the Ministry of Agriculture headed by Iakovlev. But Vavilov's very positive judgments of Lysenko's vernalization work on several occasions in late 1931 and early 1932 suggest that it was also the other way around: Vavilov's praise helped convince Iakovlev of the high value of Lysenko's work.

In May 1932 Vavilov visited the institute in Odessa. He saw the plots with vernalization experiments, and with Lysenko as guide, he inspected trial fields of cooperating collective farms and sovkhozes. In a letter to his vice director at VIR, Vavilov wrote enthusiastically: "Lysenko's work is remarkable and forces us to take a different view of many things. It is necessary that the World Collection is worked through with vernalization."[5]

As Vavilov left for the international genetics congress in the

From Problems of Plant Breeding to Controversy in Genetics | 161

United States in August 1932, he was worrying about the situation at his institute in Leningrad. In a letter to his vice director he went through various problems that might need attention. Among other things, Vavilov commented on the physiologists' resistance to vernalization. The internal disharmony among the physiologists, as well as his own disagreements with some of them, troubled Vavilov. An agreement had at least been reached that there would be collective schooling in vernalization during the winter. "I intend to learn vernalization myself." And he had already come to terms with Maksimov, wrote Vavilov.[6]

In Vavilov's lecture at the international congress of genetics, Lysenko had a prominent place. The lecture was called "The Process of Evolution in Cultivated Plants" and consisted mainly of a survey of results and projects in Soviet agricultural plant research. Lysenko's "remarkable discovery" of vernalization was particularly promising. Vavilov repeated in concise form to the international audience the claims given at various occasions at home during the preceding year:

> The remarkable discovery recently made by T. D. Lysenko of Odessa opens enormous new possibilities to plant breeders and plant geneticists of mastering individual variation. He found simple physiological methods of shortening the period of growth, of transforming winter varieties into spring ones and late varieties into early ones by inducing processes of fermentation in seeds before sowing them. Lysenko's methods make it possible to shift the phases of plant development by mere treatment of the seed itself. The essence of these methods, which are specific for different plants and different variety groups, consists in the action upon the seed of definite combinations of darkness (photoperiodism), temperature, and humidity. This discovery enables us to utilize in our climate for breeding and genetic work tropical and subtropical varieties, which practically amounts to moving the southern flora northward. This creates the possibility of widening the scope of breeding and genetic work to an unprecedented extent, allowing the crossing of varieties requiring entirely different periods of vegetation. (Vavilov 1932a, 340)

This address confirms that the uses of vernalization in plant breeding was the main reason for Vavilov's interest in Lysenko's work.

I have shown how Vavilov's interest in vernalization grew in 1931–1932. His hope for help in exploiting the World Collection is expressed in a lecture called "Problems of Selection in the USSR" given at the Leningrad "house of scholars" in April 1933: "Selection and genetics last year unexpectedly received help from physiology." With "the comparatively simple technique of vernalization," many southern varieties produced "normal harvest in more northern regions where they normally cannot ripen." For instance, wheat varieties from North Africa and the south of Spain "headed normally and gave beautiful grain" in Saratov. These discoveries showed that the value of vernalization had been underestimated. Until now we have "attributed little value to the method of vernalization."[7]

In his survey of progress in Soviet agricultural plant science during the years 1930–1933, Vavilov also wrote as if he accepted that the vernalization of spring wheat had led to increased yields: "The vernalization that was carried out in the years 1932–1933 on the fields of *kolkhozes* and *sovkhozes* gave positive results in the sense of increased yields." But the important thing for Vavilov was application to plant breeding. The method of vernalization would help demonstrate the real value of the World Collection: "The systematic collection of plant resources has given into the hands of Soviet plant breeders a colossal new material. The method of vernalization makes it easy to exploit this material in crossings" (Vavilov 1934a, 15).

The high evaluation of Lysenko's research gave Vavilov good reasons to help him obtain formal academic recognition. In March 1933 Vavilov successfully proposed Lysenko for the scientific prize of the Council of Peoples Commissars.[8] First and second prizes were given to an evolutionary biologist and a chemist. Lysenko received the third prize.[9] In February 1934 Vavilov proposed to the biological section of the Academy of Sciences in Moscow that Lysenko should become a corresponding member: "Although he has so far published comparatively few works, his latest work represents such a major contribution to world science that it permits us to propose him as a candidate for corresponding membership in the Soviet

Academy of Science." In support of the proposal, Vavilov mentioned not only the theory of stages and the use of vernalization in plant breeding but also its success as a practical agricultural technique: "... in principle this method has already been sufficiently worked out, so that this year vernalization of cereals and cotton will be carried out on a million hectares."[10]

LYSENKO TURNS TO BREEDING AND GENETICS

In 1935 Lysenko's ideas on plant breeding, seed production, and general genetics became an important public issue. His ideas were considered obscure and outdated by the experts in genetics. They were not taken seriously until Lysenko had obtained a position where he could significantly influence the national science policy in this area. His promotion to academician of VASKhNIL in June 1935 produced a new situation.

A paper titled "Theoretical and Practical Significance of Lysenko's Research on the Vernalization of Agricultural Plants" was published in 1933 in English in a journal of the Imperial Bureaus of Plant Breeding. The author was one of Lysenko's collaborators at Odessa, A. Favorov. He expressed a strong belief in the ongoing Soviet revolution in science: "Scientific work in our country, the country of Soviets, has been put on a new road, namely, reconstruction on the basis of new methods uniting theory with practice and planning in all fields, more especially the field of rapprochement between scientific institutions and the masses which are building up a new industry" (Favorov 1933, 10). The idea of a new socialist science was a source of legitimation that supplied many easy answers to scientific critics. Since a methodological revolution had occurred, the old standards of proof and testing were to a large extent irrelevant. They represented an obsolete academic culture that illegitimately isolated science from practice and from the masses.

The paper was not weighed down by Marxist theory or terminology, however. The term "dialectical materialism," for instance, was used only once. The official doctrines about unity of theory and practice and practical efficiency as the superior criterion of valuable science were central to the argument. But they were stated in a form that could appeal to a capitalist entrepreneur as well as to a communist boss. It was an attitude and ideology of science not unlike that of Vavilov.

Favorov's article used the official Soviet doctrines systematically to bolster the conclusions of Lysenko and undermine the views of his opponents. Nikolai Maksimov's explanation of vernalization and photoperiodic phenomena in terms of an antagonism between the generative and vegetative state was faulted for its lack of "close relation with human tactics" (10). Hormone theories of development, referring to a chemical stimulus, were faulted as mechanistic. They represented an "undue simplification" which obscures the "the qualitative changes in cells and tissues during the ontogeny of a plant organism." Because of a correct method, Lysenko, on the other hand, had been able to establish his results "by direct experiment" and without analytical tools like razor, microscope, or chemical analyses (11).

In Favorov's vague but suggestive criticism of classical genetics there was a holistic tendency that fit his skeptical attitude toward abstract scientific theorizing. The traditional concept of the gene and its relation to phenotypic characters necessitated a "formalistic treatment" and unavoidably led geneticists to a "mechanistic interpretation." Lysenko's work, on the other hand, promoted "more concrete conceptions on the question of origin of characters" and would thus "eliminate the gulf that lies between the gene and the character and sometimes introduces a great deal of confusion in the understanding and practical utilization of the phenomena of heredity and variation" (Favorov 1933, 12). This paper presented a program for radical change in genetic science using studies of individual development (embryology) as the main tool. This far Lysenko and his followers were in line with many critics of classical genetics in the West as well as in the Soviet Union.[11]

In early 1934 Lysenko himself published a paper called "Physiology of Plant Development and Selection Work," which included critical comments on classical genetics. The article was based on a lecture at an All-Union Congress for Seed Growers. Lysenko was dissatisfied with the tendency to apply vernalization eclectically as a mere technique, without taking into account the underlying theory. He appealed to the Stalinist practice criterion: Many researchers apparently had the impression that he was against all theory, and in particular against genetics and selection, said Lysenko. This was due to misunderstanding of his theory of stages. He was for the kind of theory that, according to comrade Stalin, "shall give the practical man power of orientation, clarity of purpose, confidence in work and belief in victory" (Lysenko 1934a, 22).

Lysenko complained that breeders and geneticists did not sufficiently appreciate the role of the environment. They pay lip service to the principle that all "characters, properties, and qualities" of the plant are results of its individual development and affected by the environment. But they make the mistake of letting characters that are always the same depend only on heredity. It is just as wrong, said Lysenko, to hold that development does not depend upon the environment as to hold that it does not "depend on the hereditary basis" (Lysenko 1934a, 24). His claims could stand as a paraphrase of Johannsen's formula, "heredity + environment = phenotype," and thus represent the very core idea of classical genetics. Nevertheless, Lysenko made them into an argument against classical genetics as the basis of plant breeding, an argument that impressed many people outside the circle of genetic specialists.

When Lysenko used the terminology "genotype" and "phenotype," he did not think in Johannsen's terms. He did not conceive the genotype as composed of stable entities that are transferred unchanged from one generation to the next. For Lysenko, the genotype was subjected to continuous change, like the organism is during its development from germ to maturity. Three decades earlier there had been a big fight between biometricians and Mendelians over continuous versus discontinuous change of biological heredity. Lysenko held on to the biometric belief in continuous change. This provided an opening for immediate and directed influence of the environment on heredity, or, in other words, for the inheritance of acquired characters. Lysenko also held that classical genetics, at least "formal genetics," was really concerned "not with genes but with phenes" (28).[12] Presumably Lysenko misinterpreted classical genetics more from a lack of understanding than from ill will. We have seen that similar objections about classical genetics' disregard of the environment were raised by Soviet biologists who had much better knowledge of genetics than Lysenko.

The main concern of Lysenko's 1934 paper was an experiment in wheat breeding that he had started in 1933. I have described how the technique of vernalization had become a major instrument for the efforts to make plant breeding more effective. Lysenko now wanted to show that not only the technique of vernalization but also his theory of stages had important implications. He pointed out that late ripening could be linked to different stages in the development of a plant. If two varieties were late for different reasons, that is, due to properties connected with different stages, then the offspring obtained by crossing them could well be early ripening. Lysenko's proposal for the breeding of new early-ripening varieties was that in the choice of pairs one should aim to combine a short temperature stage with a short photo stage (Lysenko 1934, 27). Analysis by vernalization would help to pick the right kinds of plants. It was necessary to distinguish not simply between late and early ripening, but quantitatively to determine differences in the duration of each particular stage of development.

Such an experiment had already been started, in January 1933, announced Lysenko. With the help of modern techniques, greenhouses and vernalization, the fourth generation had been reached after only one year. And individuals had been discovered that ripened earlier than both the original parents, just like predicted. Lysenko thus maintained that his theory of stages had solved a problem with which classical genetics had struggled in vain. The reason for its failure was that classical genetics had not properly distinguished the different genetic bases for the character of late ripening, argued Lysenko (1934, 27–28).

When this experiment was proposed in January 1933 it met with skepticism from several "not unknown geneticists and breeders," said Lysenko (1934, 27). From their perspective of classical genetics they could not see anything new and interesting in his proposal. The experiment that Lysenko started in January 1933 was apparently conceived as a kind of crucial experiment to demonstrate the truth of his theory over classical genetics. He probably hoped for a similar effect as in 1929, when the experiment on his father's farm had started a process that forced the plant physiologists to take his theories about vernalization seriously.

Lysenko's tendency to prefer simplistic and speculative theorizing to thorough empirical research soon precipitated Vavilov's irritation. In 1934 Lysenko gave a lecture at VIR titled "Plant Physiology and Selection Work."[13] In the discussion Vavilov complained that Lysenko had used most of the time to teach elementary genetics. Not only was it elementary, it was also confused. Vavilov took some time to clarify and point out the shortcomings of Lysenko's ideas on hybridization, genotype, and phenotype. "If he had only opened Johannsen's treatise, he would have found a brilliant exposition of the theory of genotype and phenotype." Lysenko interrupted with a protest that there was scientific disagreement. Vavilov replied by deploring that Lysenko was neglecting serious work to develop more effective methods in plant breeding. "From you, comrade Lysenko, we expect concrete work on these matters."[14] Vavilov no doubt saw the serious flaws in Lysenko's scientific competence, but for a couple of years he continued to be highly restrained in his public criticism.

In the early summer of 1935 Lysenko, with Prezent as coauthor, published a small book that presented views on plant breeding built on the theory of stages. The criticisms of traditional plant breeding and classical genetics were sharper and more systematic than before. A central part of the argument was the experiment in wheat breeding that Lysenko had started in January 1933. The experiment was succeeding and the theory behind it had thus been proved, they argued.

Lysenko and Prezent claimed that new laws of inheritance had been found that made it possible to work with a much smaller number of plants than was prescribed by accepted methods. If true, this discovery would greatly increase the productivity of plant breeding. Already in the first generation of hybrid offspring, the plants that had the potential of producing varieties with early ripening would distinguish themselves, according to Lysenko and Prezent. First, they maintained that there would regularly occur first-generation hybrids ripening earlier than, or at least as early as, the earliest of the parents (Lysenko and Prezent 1935, 39). Second, they argued that later generations of offspring could not produce forms that would ripen earlier than the first (45). These two rules implied that already in the first generation of offspring one could, with a high degree of certainty, pick a small number of plants as those with promise, namely, the ones that were clearly earlier than both parents. In all following generations one could continue to pick the earliest ones and throw away the rest. According to generally accepted principles, the situation was much more complex. For instance, it would be the exception rather than the rule that the new, extra-early forms appeared already in the first hybrid generation. Through segregation, late forms would continue to give rise to early ones for many generations. It was therefore necessary to continue growing the descendants of a plant for several generations before one could say with some degree of confidence that it would never produce an interesting early variety.

The two principles maintained by Lysenko and Prezent represented an important development from the lecture in January 1934. Lysenko and Prezent now explicitly contradicted important results of classical genetics. It was no longer possible to say that Lysenko's way of applying the theory of stages to plant breeding was fully compatible with classical genetics.

The weakness of Lysenko's two laws was that they presupposed simplified conditions that are not found in the real world. For instance, there is usually more than one gene pair involved in the determination of a single property such as a short temperature range or a short photo stage. Furthermore, it will not always be the case that early ripening is dominant over late, as Lysenko assumed. It may be the other way around, or dominance may be incomplete. Generally accepted genetic knowledge implied that Lysenko's laws were unlikely to have any broad validity. And the evidence that he had produced gave them little support.

However, it is hard to find basic principles that are absolutely incompatible with classical genetics in the 1935 booklet of Lysenko and Prezent. It does not, for instance, contain any explicit thesis that acquired characters can be inherited. But by using Ivan Michurin as a theoretical authority, with his predilection for the environment as an explanatory factor, the ground was prepared for the development of Lamarckian ideas. One could say that Lysenko pushed aside the distinction between phenotype and genotype and returned to earlier nineteenth-century ideas about heredity. These he supported with quotations from Michurin, Timiriazev, and Darwin and by playing on the intuitive notions of agricultural practitioners.

Lysenko and Prezent treated genotype or heredity as a property of the whole organism. Like other properties of the whole organism, it was conceived as essentially dependent upon the environment. The idea of heredity as linked to a stable and well-defined physical structure or condition in a part of the organism, something that is naturally transferred unchanged from one generation to the next, was absent. On the contrary, the genotype was described as a *possibility* for development in different directions, dependent upon environmental conditions (Lysenko and Prezent 1935, 10).

THE JUNE 1935 MEETING IN ODESSA

The meeting of VASKhNIL in Odessa in June 1935 was primarily devoted to the work of Lysenko and his associates. Lysenko's address had two main messages, both more or less in conflict with classical genetics. First, there was the selection of pairs for hybridization. Second, Lysenko made a proposal for seed growing based on ideas even more clearly in conflict with classical genetics. He argued for artificial cross-fertilization of purebred varieties of self-fertilizing grains in order to counteract degeneration.

An extensive report in a leading agricultural journal gave the impression that the academy found Lysenko's experiments and arguments convincing, accepted his theories, and recommended his practical proposals. Concerning the theory of stages, "there are no diverging opinions in the scientific world," claimed the report, and its importance was stressed even by the most enduring skeptics (Gatovskii 1935, 206). But other sources reveal substantial criticism. In his concluding talk, Lysenko tried to answer some of the objections. For instance, Vavilov had criticized the "law" that early ripening was always dominant over late. He claimed that a survey of the international literature showed that this relationship varied. According to Japanese investigations, late ripening in rice was dominant in all cases that had been investigated. Lysenko replied that ten different crossings of rice at his institute had given the opposite result (Lysenko, 1935a, 57–58). A wheat breeder, Lepin, claimed that in his experiments the first-generation offspring partly ripened later than both parents. Lysenko answered that the difference was too small to be significant and that two different years had been compared and the conclusion therefore not tenable (56–57).

Once again Lysenko revealed his lack of scientific training and tendency to dogmatism when he had become convinced about the truth of a theory. He showed little ability to critically weigh evidence for and against according to the established methodological norms of experimental science. Lysenko mocked the accommodating character of Vavilov's criticism: He is so full of reservations that he becomes self-contradictory:

> Academician Vavilov in his contribution declared that he is 90 percent in agreement with our proposals and endorses our method of selection. He has refuted the theory of inbreeding in a logical and convincing way, and still he only accepts this conclusion 90 percent. But the remaining 10 percent prevents him from completely seeing through the theoretical falsity of inbreeding. Without such a complete understanding it is impossible to concur in our proposal on seed production for self-fertilizers (Lysenko 1935a, 59).

Lysenko's proposal for invigoration of standard varieties through crossbreeding was linked to a rejection of Johannsen's theory of pure lines. According to Johannsen, pure lines could change only through mutation or hybridization, but experience showed them to be considerably less stable. Pure lines also changed because of a "continuous segregation of the hereditary basis," claimed Lysenko. Degeneration in pure lines of self-fertilizers was to be expected in accordance with the well-known harmful effects of inbreeding in general (Gatovskii 1935, 209).

Lysenko claimed that selected varieties of rye, which is a cross-fertilizer, are generally more stable than selected varieties of wheat. And he speculated that a cross-fertilizing wheat variety would be more stable and more productive than the present self-fertilizing varieties. Lacking such a cross-fertilizing wheat he proposed that artificial cross-fertilization be introduced in seed production. Giving the egg cell an opportunity to choose fertilization by pollen from another plant was in Lysenko's terminology analogous to opening the possibility of "marriage for love," which was advantageous according to the theories of Darwin as well as Michurin. Technically, the process was to be executed by hand with the help of a pair of pincers, and Lysenko envisaged the mobilization of an army of peasant-scientists from the so-called hut-laboratories to implement this scheme (Lysenko 1935a, 47–54).

At the VASKhNIL meeting in Odessa, Lysenko still could be taken to argue within a conceptual framework that did not exclude classical genetics. He talked about genes, homozygotes, and heterozygotes. He discussed pure lines, mutations, and recombination of genes. He did not set out a radically different paradigm even if he criticized and rejected some of the principles of classical genetics as it had been formulated by Johannsen and the Morgan school. As we have seen, such criticisms did not disqualify Lysenko as a biologist. They were widespread in the international scientific community in the 1930s and cannot by themselves count as evidence for Lysenko's scientific incompetence.

Three weeks after the excursion to Odessa, Lysenko's ideas were subjected to sharp criticism at a meeting of the presidium of VASKhNIL.[15] Lysenko was not present himself. Leading critics were vice presidents Zavadovskii and Meister. Attacks on Lysenko were also made by Konstantinov, an experienced and productive plant breeder, and Lapin, head of the seed production section of the VASKhNIL secretariat. Vavilov was critical in some respects, but he also acted as the main defender of Lysenko with inarticulate support from Eikhfel'd.

In a long introductory talk Vavilov reported on the Odessa meeting. Once more he emphasized the great value of vernalization and also took a generous attitude toward Lysenko's new methods of plant breeding. Vavilov left no doubt concerning Lysenko's new genetic ideas, however. They were contrary to established genetic theory as well as accumulated facts of inheritance.

Lapin argued that in a number of instances the vernalization of seed grain appeared to have had a negative rather than a positive effect, and that its efficiency as a method to increase yields was still in doubt.[16] Meister was particularly sharp in his polemics. He maintained that Soviet breeding, including the successful work of Lysenko, was built on the principles of classical genetics. The numerous successes, especially in recent years, gave ample evidence in favor of classical genetics. The criticisms of Lysenko and Prezent were unfounded and particularly inappropriate now when the Soviet Union had taken on the next International Congress of Genetics. It would make a bad impression abroad if a vulgar "marketplace" criticism (*kritika rynochnogo kharaktera*) of genetics by Soviet academicians was to be read in newspapers "sold in every kiosk," said Meister.[17]

As the discussion continued, Zavadovskii attacked Prezent for his unscientific "Darwinism,"[18] and Konstantinov maintained that so far the vernalization of seed grain had not been tested by a proper scientific method. "One can talk with closed eyes about the usefulness of vernalization, but there are no data," he said.[19] Bondarenko, the newly installed academic secretary of VASKhNIL, stressed that at least three years of successful trials were needed before Lysenko could rightly claim to have produced a new valuable variety.[20] And Muralov, who had just become president, wondered whether Lysenko in his laudable boldness had crossed the border to unhealthy extremism.[21]

Vavilov replied with a passioned defense both of Lysenko and of his own role as president of the Odessa meeting. This meeting had been organized mainly to evaluate the work on vernalization, and every academician who spoke had been positive in his or her judgment. Lapin, for instance, had not mentioned his doubts in Odessa. Vavilov spoke with conviction about the practical usefulness of vernalization. He had reviewed Lysenko's results on behalf of the Commissariat of Agriculture starting in 1931. In *kolkhozes* and *sovkhozes*, as well as in the work of Lysenko, he observed "definite mass indicators" that were "positive in particular with vernalized wheat." Vavilov also argued that vernalization would permit the growing of high-yielding English varieties in the southeastern region beyond the Volga. These varieties have poor baking qualities, he admitted, "but they are needed for biscuits, etc." "To me," said Vavilov with emphasis, "it is quite clear, I am absolutely convinced, that in agrotechnology vernalization is a great achievement." And he added that new possibilities were opening up in improving the quality of products. For instance, it had turned out that the length of fibers in flax was increased by vernalization.

Finally, Vavilov mentioned the application to breeding. Until recently, the World Collection had not been much used, he admitted. But during the last few years it had become "accessible through the use of vernalization." Vavilov even quoted the well-known British cotton breeder S. C. Harland as saying that "vernalization is the third greatest achievement in world science."[22] Lysenko's selection of parent plants for hybridization was another important contribution, and summer planting of potatoes to avoid degeneration was a new interesting discovery, in Vavilov's opinion.

Even Lysenko's claim to have bred a new early variety of wheat in record time was positively evaluated by Vavilov. Lysenko had made a good choice of parents and obtained a striking result. Of course, he had not yet produced a new variety for practical farming. "We were singing the praise of the method," said Vavilov, not of "an accidental new variety." The "psychological" situation at Odessa necessitated a statement on this topic, Vavilov explained. If Meister and some others speaking at this meeting had also been in Odessa, they would have understood how hard it was to find a way to confront youthful enthusiasm and "impatience" with tactful corrections. In this atmosphere "the nervousness" of the main speaker, that is, Lysenko, was "increased," said Vavilov diplomatically. With the help of Eikhfel'd, he had obtained a promise from Lysenko that he would cross out the strongest expressions from this oral presentation. Vavilov also reminded the meeting that worldwide a critical revision of the principles of genetics was taking place.[23]

Thus in July 1935 Vavilov was still full of enthusiasm for vernalization, and he valiantly defended Lysenko against criticism, belittling his faults and upgrading his achievements. Apparently Vavilov's joy over the prospects for the exploitation of the World Collection had somewhat clouded his judgment on other issues.

The meeting of the presidium of VASKhNIL on July 17, 1935, marked the start of the virulent public controversy over genetics that culminated one and a half years later, in December 1936, with the congress on "Disputed Questions of Genetics and Selection." The resolution passed by the presidium at the end of this meeting stated that Vavilov, Meister, and Konstantinov were to publish discussion papers to clarify the disputed questions of breeding, genetics, and seed production.[24] There was an obvious need to inform the public about the counterarguments to Lysenko. It was not acceptable that national newspapers like *Izvestia* simply took Lysenko's claims at face value by publishing his article "Rejuvenating the Seed," followed by a brief note which simply repeated his main claims: Lysenko had succeeded in producing new, superior varieties of spring wheat in only two and a half years, and he had found a "very simple way" to renew the main wheat varieties in two to three years by cross-fertilization.[25]

NEW WHEATS IN RECORD TIME

Lysenko continued to push his new ideas on breeding and seed growing after the Odessa meeting in June 1935. The first issue of Lysenko's new journal, *Iarovizatsiia*, opened by reproducing an official declaration from the Odessa institute to the agricultural authorities.

It was announced that four new early-ripening wheat varieties had been produced in record time and that the new method of wheat breeding had thus been proved right.[26] At the same time

Lysenko's team had fulfilled the goals of the government directive of August 1931 to cut the time for breeding new varieties of grain from ten to twelve years down to four to five years.

> With your support our promise to breed in two and a half years, through hybridization, a variety of spring wheat for the Odessa region which is earlier and more productive than the regional variety 'Liutestsens 062', has been fulfilled.

According to the accompanying data, the new varieties gave large increases of 20 to 40 percent in yields compared to the standard varieties of the region. The declaration was dated July 25. It thus followed up and further confirmed the results that had been presented at the Odessa meeting at the end of June. Criticism here and later had not led to more modest claims.

The method of renewing wheat varieties by artificial cross-fertilization was also followed up. Lysenko proposed that this new method should be applied on a large scale without previous practical testing. "Our theoretical premises (which have not yet been verified in practice) give us reason to hope for great practical effectiveness in renewing the seed of self-fertilizing varieties." Once more, theory and practice were fused in a way that made effective testing and evaluation difficult.

Meister followed up his sharp criticism at the July meeting of the VASKhNIL presidium in a paper published in October 1935. He noted Lysenko's new early-ripening forms of wheat, produced in two and a half years, as a very interesting result. But these were not yet new varieties in the full sense, warned Meister. A much more extensive program of testing is needed before a new variety can be recommended for practical agriculture. So many other demands than early ripening have to be satisfied. The basic principle of selection is "many are called upon but few are chosen."

What really infuriated Meister was the tendentious account that Lysenko and Prezent gave of classical genetics, in particular their irresponsible claims that early ripening is a dominant property and that pure lines necessarily degenerate with self-fertilization. "The position that comrades Prezent and Lysenko have taken with respect to contemporary science is incomprehensible and does not agree with the philosophy of the proletariat," declared Meister. The positive value of the theory of stages is no reason for ridiculing research on chromosomes (Meister 1935a, 13). From a practical point of view, the worst offense was the heedless attitude toward the existing system for seed production. Meister commented on a newspaper article reporting on Lysenko's talk at the Odessa meeting: One cannot forbid Prezent and Lysenko to make statements on what they wish. "But it is completely incomprehensible that the official organ of the Soviet Ministry of Agriculture publishes an article that disorganizes government institutions of selection and seed growing" (Meister 1935a, 16–17).[27]

These must have been hard words to swallow for a man with Lysenko's ambitions and quickly rising status. What Meister expressed was presumably a widespread opinion among established specialists. He was one of the very few who had the social and ideological as well as scientific credentials to make this kind of politically loaded criticism in public.

Perhaps Meister came to regret these strong words against Lysenko. He later became much more accommodating. It is interesting to note that in this case, Lysenko's reply, printed together with Meister's criticism, was sober and low-keyed, in fact, a retreat from his earlier brash claims. Lysenko repeated at considerable length the theoretical views about degeneration by inbreeding. What he had proposed was practical experiments with artificial cross-fertilization, explained Lysenko. He could not understand how such experiments could represent a threat to the existing rules for seed production (Lysenko 1935b, 21).

Wheat growing and the breeding of cattle and sheep were the two main themes of the October 1935 session of the VASKhNIL conference held in Saratov. This was the first general scientific conference of the reorganized VASKhNIL. Besides the vernalization of seed grain,[28] the breeding of new wheat varieties was also a main topic.

Meister lectured on results and perspectives in Soviet wheat breeding. He stressed the importance of the theory of stages for the

From Problems of Plant Breeding to Controversy in Genetics | 177

selection of pairs (Meister 1935b, 142). He also acknowledged that the tempo in breeding had to be increased and that Lysenko's production of "new forms of wheat" in record time was an interesting example of applying the techniques of modern plant physiology (147). But at the same time, Meister made it quite clear that classical genetics must remain the theoretical basis of breeding. A scientifically based systematic breeding of plants "was made possible only after the discovery of the botanist Johannsen." His work on the pure lines was basic both to simple selection and to breeding through hybridization (Meister 1935b, 142).

Meister thus maintained a clear differentiation between Lysenko's plant physiology and his genetics. While he acknowledged the important contributions of the former, he did not even bother to discuss the specific contents of the latter. Meister's defense of classical genetics was probably also addressed to Muralov, who had given a negative evaluation of the practical results of classical genetics in his introductory lecture at the meeting. He had said that until this time, "genetics has developed completely independently and completely cut off from practical tasks of selection. Genetics has not given any directions for the choice of pairs of parents for hybridization" (Muralov 1935a, 16). However, Muralov had not explicitly mentioned Lysenko's genetics as an alternative, and he did not repeat the negative evaluation of classical genetics in his summing up of the conference (Muralov 1935d).

In Vavilov's survey of wheat varieties from the whole world and their exploitation in Soviet breeding, Lysenko was not mentioned (Vavilov 1935b). His attitude toward Lysenko had probably changed since the summer. Neither did Lisitsyn in his lecture on seed production mention Lysenko. The proposal of intravariety hybridization to prevent degeneration was simply ignored (Lisitsyn 1935).

But how much did the experts' rejection of Lysenko's new ideas on breeding and genetics impress the political leadership of VASKhNIL and the political establishment outside? There persisted a strong belief that a new unity of theory and practice would produce a methodological revolution and bring Soviet science ahead of science in the capitalist countries. In his report on the results of the

October conference, Muralov pointed to the Stakhanov movement and its implications for science. Not only in industry but also in science, "a new tempo and new methods" were needed (Muralov 1935d, 109). That new and better methods could be found appeared obvious to the politicians. The problem was only to choose the right ones. Here they looked to the practice criterion for help. "Socialist competition" was a way to organize science that made this criterion more effective.

DIALECTICAL MATERIALISM IN GENETICS

Through a gradually radicalized critique of classical genetics, Lysenko arrived at an explicit preference for a neo-Lamarckian over a neo-Darwinian view. The possibility of changing the heredity of an organism through "education" by the environment became a basic principle in his genetic thinking. Now, in the mid-1930s, Lysenko attempted to resurrect what amounted to the central idea of neo-Lamarckism. The Lysenkoists, with considerable success, used "dialectical" rhetoric to support Lamarckian claims.

But Lysenko was by no means alone in promoting a dialectical materialist critique of classical genetics. This was a broad movement in Soviet biology at the time. Lysenko was a fellow traveler rather than a pioneer of this movement. There were traditional biologists oriented toward the ecology, systematics. We have mentioned Vladimir Komarov and Boris Keller. There were plant physiologist like Vladimir Liubimenko. But even among geneticists, this was a strong trend. Significant examples that we have already mentioned are Sapegin and Meister. Vavilov also believed in the need for a radical revision of classical genetics. In his lively dispute with Meister over the virtues of Lysenko's research in July 1935, Vavilov pointed to Sapegin's accusation of 1932 against the Morgan school for its lack of dialectical depth. Lysenko's proposals had received so much attention because they touched on serious problems with classical genetic theory.[29]

During the last half of 1936 the journal *Socialist Reconstruction of*

Agriculture carried a number of discussion articles in preparation for the VASKhNIL genetics congress in December of that year. Just before the congress, Keller published a paper on "Genetics and Evolution." He argued from an ecological and natural history point of view that the geneticists' perspective was narrowly experimental. They had forgotten the historical dimension and had not grasped the evolutionary worldview, claimed Keller, who quoted Kliment Timiriazev as his authority (Keller 1936, 24). Keller based his ideas of genetics in part on speculation about the very earliest stages in the history of life, before the first cell formed. He held that a number of the basic features of life had been formed in this period.

The last section of Keller's article discussed how phenotypic modifications could affect heredity. He focused on Michurin's theory of "mentors" and Lysenko's theory of stages. The latter had inspired Keller to rebuild his own special discipline, plant ecology, on a "new dynamic basis" that took into consideration the "deep dialectic unity of environment and plant." One must not simply reject Lamarckism as a whole, argued Keller, but make use of its valuable elements. Both Friedrich Engels and Michurin had believed in the inheritance of acquired characters. And Keller saw no reason why there could not be some truth in this idea, in the sense of the transformation of modifications into mutations. This was a question that needed further investigation (Keller 1936, 30–32).

The idea that environmental modifications could be transformed into genetic mutations had appeal also among professional geneticists. It was by no means an outlandish and a priori impossible hypothesis. In fact, it was a main working hypothesis for at least one geneticist who played a central role in the later stages of the genetics controversy, namely, Anton Zhebrak.

Anton R. Zhebrak (1901–1965) was, like Lysenko, a peasant's son. But he had taken a regular education at the Timiriazev academy, and he was politically active, having joined the Communist Party during the civil war. Zhebrak graduated from the Timiriazev academy in 1925 and the Institute of Red Professors in 1929. He had also studied at Columbia University in New York in the group of students who studied under Leslie Dunn and at the California

Institute of Technology with T. H. Morgan in 1930–1931. From 1932 he worked at the Timiriazev academy, becoming professor and head of the genetics department in 1935.

The August 1936 issue of *Socialist Reconstruction of Agriculture* contained two articles defending genetics against the Lysenkoist attacks. The journal presented the two articles as issuing from a discussion about the Seventh International Congress of Genetics planned for Moscow in the following summer. Both authors professed loyalty to dialectical materialism but were quite different in their attitude toward Lamarckian ideas. The first paper, by M. Zavadovskii, vice president of VASKhNIL, simply dismissed them as uninteresting, while the second, by Zhebrak, took Lamarckism seriously as an important theoretical possibility and interesting working hypothesis. There was also a difference in their willingness to accept a principal difference between theoretical and applied science.

As a budding embryologist, Mikhail M. Zavadovskii (1891–1957) had been one of the young turks of Serebrovskii in the attack on Lamarckism in the late 1920s. Serebrovskii had then opposed the doctrine of two sciences by arguing that capitalist science could well produce objective results valid for a socialist society. Zavadovskii now argued for a distinction between theoretical and applied science. Fundamental scientific knowledge was politically neutral, but its application ought to follow socialist principles.

Zavadovskii directed his sharpest criticism against Prezent and in particular against his argument for a special "historical" dimension in the method of biology as opposed to the sciences of nonliving nature. Prezent had argued that classical genetics and experimental embryology were too mechanistic in their approach. They excluded the "historical-biological aspects" of Darwin's method (Prezent 1936). Zavadovskii replied that Prezent's appeal to a special historical type of explanation in biology was simply built on a conflation of ontogenesis and phylogenesis. He lacked a proper distinction between *individual development*, which is the theme of embryology, and *evolution of species*, which is the theme of the theory of evolution (Zavadovskii 1936, 88–90). Zavadovskii did not mince his words: Prezent did not know what he was talking about, "his activity is in fact obscurantist" (95).

Lysenko was treated much more mildly. Zavadovskii did not question the value of his practical agricultural results and gave full recognition to his theoretical contributions in plant physiology. Lysenko had been misled by Prezent to make "careless generalizations" in theoretical questions (Zavadovskii 1936, 96). The fundamental mistake was that Lysenko considered himself a geneticist. His experiments on vernalization and his theory of stages belonged to the study of the development of individual organisms, ontogenesis. They did not properly engage in the study of hereditary differences between successive generations, genetics, or the study of the forces that mold species, evolutionary theory. During the preceding twenty-five years, a differentiation had developed between these three levels or types of investigations, a differentiation that Prezent and Lysenko had not grasped. Lysenko was studying individual development, not genetics or evolutionary change (Zavadovskii 1936, 86–88).

Zavadovskii admitted that Lysenko's criticism in practical matters was not without some justification. It is correct that genetics is separated from practice, wrote Zavadovskii, but as a theoretical discipline, it should be so. The mistake is that leading geneticists have not seen clearly enough the difference between theoretical and applied science. They have overestimated the ripeness of genetics as a science and created completely unrealistic expectations about the practical technological results that can be achieved at the present stage. "We are therefore witnessing a natural disappointment," Zavadovskii explained (1936, 96).

Zavadovskii quite explicitly built his attempt to straighten out the confusion in methodology and science policy on a distinction between theoretical science, the "university type," as he called it, and applied science, that is, plant and animal "technology." These are sciences on different levels. Theoretical science studies "forms of movement objectively existing in nature," while applied science is a means to reach human goals. Like physics and chemistry, genetics can be useful for practical work. But it is also possible to do practical work without much knowledge of basic theories in biology. However, this possibility does not deprive science of its right to exist, wrote Zavadovskii (1936, 95).

Zavadovskii claimed to build on dialectical materialism, referring to authorities such as Engels. But his views conflicted with the popular doctrine of unity between theory and practice, and it was easy to stamp him as representing obsolete academic traditions. It was not accidental that during the years 1936–1937 Lysenko's journal *Iarovizatsiia* printed just one article defending classical genetics, namely, this one by Zavadovskii.

Zhebrak, in his article titled "Some Contemporary Problems of Genetics," maintained that the basic principles of classical genetics were fully compatible with dialectical materialism. But he admitted that there existed nondialectical, idealist, and mechanistic tendencies within the discipline. His ambition was to construct "a third point of view," which would unite the internal and the external perspectives, epigenesis and preformation, Lamarck and August Weismann, and also "add something new" (Zhebrak 1936a, 104). The German biologist Weismann's distinction between germplasm and soma (body), which Johannsen had developed into genotype and phenotype, was a fundamental advance in the history of genetics. But their thinking was tainted by a dualistic philosophy that tended to see genes or genotype as isolated from the body and completely unaffected by its changes, argued Zhebrak. Like Keller, he saw the inheritance of "modifications," that is, variations owing to environmental differences, as crucial. He thought that in some sense this had to be possible if the objectionable dualism was to be overcome (Zhebrak 1936a, 103–104). He stressed that the *relative* stability of the genes was a fundamental principle in genetics, as Johannsen had maintained. But at the same time the genes were changing. We "have no grounds for maintaining that the genes of the Egyptian pharaohs are exactly like those of their contemporary descendants," Zhebrak argued (109).

Thus Zhebrak took the criticism that classical genetics was not dialectical much more seriously than Zavadovskii. He was not satisfied with Zavadovskii's simple solution through the distinction between the evolution of the individual and the species. Zhebrak felt that the concept of historical change was so fundamental that it had to be incorporated already in the causal explanations of genetics and individual development. While Zavadovskii did not worry much about the consequences of an analytic approach that would use only knowledge about the parts and their relations to explain the whole, Zhebrak stressed the fundamental significance of the whole (Zhebrak 1936a, 103). The difference was not that Zavadovskii was lacking interest in whole genomes, organisms, species, ecosytems, and so on. He simply felt that the traditional method had an adequate grasp of the relationship between the parts and the whole.

Zhebrak's criticism of Lysenko's special ideas on plant genetics and breeding methods—dominance, segregation, inbreeding, and so on—was clear but low-keyed. He discussed Lysenko's ideas seriously, interpreting them as reasonable objections to traditional views. But he found, on balance, that established ideas and practices were preferable. Zhebrak wanted in a friendly and constructive way to demonstrate to Lysenko the important achievements of modern genetics. "Our criticism is the criticism of friends, who wish to see comrade Lysenko reach a higher theoretical position, appreciating the achievements of contemporary genetics" (Zhebrak 1936a, 121–22). Zhebrak was also more friendly than Zavadovskii by not attacking Prezent as the evil spirit misleading Lysenko.

The theory of stages and Lysenko's claims about vernalization were uncritically accepted by Zhebrak. He expressed full agreement about the usefulness of vernalizing seed grain, for instance. "All these are issues on which there is not the least doubt among the representatives of genetics that I know," he asserted (Zhebrak 1936a, 112). It is hard to understand that Zhebrak did not here speak against better knowledge, or was at least fudging the issue. For several years he had been working at the Timiriazev academy, where Lisitsyn was his colleague. He must have been aware that many leading agricultural scientists doubted Lysenko's claims. His statement could be a result of wishful revolutionary thinking or of political pressure and threats. It could also be an indication that he worked at a certain distance from practical agricultural problems, like many of the other theoretically inclined geneticists and plant physiologists.

The unity of theory and practice was the main theme of a new article by Zhebrak, published in December 1936, shortly before the genetics congress. He started by insisting on this unity and warning

184 | **THE LYSENKO EFFECT**

against the separation of theoretical science from practical concerns. Nevertheless, the thrust of his argument was to give theoretical science a considerable degree of autonomy as well as authority in practical questions. Zhebrak discussed and rejected Lysenko's claims with respect to segregation in hybrids and genetic effects of grafting and inbreeding. But on one central question Zhebrak came close to Lysenko's position, namely, the inheritance of modifications. It was the reading of Engels that had inspired him to start experimental work on the inheritance of modifications in the laboratory of T. H. Morgan in America in 1930, Zhebrak explained (1936b, 87). Through these experiments, he had developed the idea that there is a mechanism by which "modifications" can direct mutations.

Zhebrak's working hypothesis was that modifications produce changes in the following generation by affecting the germ cells. At first these changes do not have the character of "mutation" but represent "a weaker change of the genotype in the form of changes in genic bonds." Zhebrak defined "genic bonds" somewhat cryptically as "bonds between generations" (1936a, 103). The accumulation of such changes in "genic bonds" over time could lead to "sharper changes of a mutational character" (Zhebrak 1936b, 8).

The abstract and speculative character of Zhebrak's article, as well as his openness to ideological argument in science, clearly comes out in his concluding sentence. After describing his genetic hypothesis, Zhebrak asserts: "To us it appears true because it agrees on the whole with the methodological conceptions of Engels and helps eliminate the dualism between phenotypical and genotypical change" (Zhebrak 1936b, 88).

VAVILOV'S ROLE

In early December 1936, just before the VASKhNIL genetics conference, a collection of relevant papers was published by VASKhNIL. These papers had been printed in various journals during the preceding two years. The collection contained papers by the main authors that have been commented on and analyzed in this chapter:

Meister (one paper), Konstantinov (one),[30] Zhebrak (two), M. Zavadovskii (two), Keller (one), and Lysenko (four). The introductory paper was by the recently deceased Michurin. But one person was conspicuously missing, namely, Vavilov.

Vavilov generally kept a low profile in the public debate over breeding and genetics in 1935–1936. In public he avoided direct and clear criticism of Lysenko's claims. He limited himself to a defense of his institute, VIR, and his own work against direct attacks. In a paper of December 1936 Vavilov answered two critics from his own institute. One was the head of VIR's department for the introduction of foreign plants. He had been involved in a public controversy with Vavilov already in 1931, and from 1933 he had been working for the secret police informing on Vavilov (Popovskii 1984, 139). The other was also a specialist on the introduction of new varieties and an informer for the secret police.

In his reply, Vavilov defended his law of homologous series. The critics interpreted it in scholastic fashion, making it into a Procrustean bed for selection and breeding work, he objected. In fact, it had been a very useful and quite flexible instrument. He willingly conceded that his early conceptions about the genetic basis of such series had been simplistic. But it was quite natural, and by no means nondialectical, that scientific concepts were subject to change (Vavilov 1936, 33–36). Vavilov called the accusation that VIR, and Soviet plant breeding generally, had neglected local varieties a "perversion of the facts." One of the great achievements of Soviet plant breeding was precisely the effective use of the local material. This was the first step of Soviet selection, asserted Vavilov (36). He also rejected accusations that the foreign varieties that had been introduced were no good. In fact, some of them had proved to be very valuable (38).

That VIR had neglected the work of Michurin was another accusation that Vavilov found offending. Quite the opposite was true. Vavilov had personally taken great interest in Michurin's work and had helped him obtain recognition and material support. Scientists from VIR had studied Michurin's work attentively, and his ideas about distant crossings had been developed to give many important results. The work of Georgii Karpechenko on doubling the chromosome number as a means to make distant hybrids more fertile had, for instance, attracted much interest at the International Congress of Genetics in Berlin in 1927. A number of new varieties of various cultivated plants had also resulted from work with distant hybridization (Vavilov 1936, 43–46). What Vavilov did not write was that among Michurin's younger collaborators, some of whom had spent time as aspirants at VIR, a strong opposition to classical genetics had developed.[31]

Two critics, A. K. Kol' and G. N. Shlykov, both on the VIR staff, each contributed a paper to the discussion collection of VASKhNIL. Vavilov's estimastion of Kol's paper was such that at one point he said he would withdraw from the academy if it was included. In Vavilov's opinion, this was not a serious scientific paper that deserved further discussion. At the later VASKhNIL *aktiv* in March 1937 the academic secretary, Margolin, severely reproached Vavilov for his threat to withdraw and thereby disrupt the work of the academy: with "your authority, your learning," Nikolai Ivanovich, you could easily have refuted the claims of Kol' and yet you threatened us with resignation. "You threatened us."[32]

This episode shows that Vavilov could be quite sharp and explicit in his criticism of the way the political leadership of VASKhNIL promoted untenable science because they were not able to judge the quality of the arguments. In this case, he also had to give. Vavilov was apparently under considerable political pressure. Toward Lysenko, he was still respectful and circumspect in his criticism.

Among the leading scientific staff of VIR, many had developed a strong antagonism toward Lysenko by late 1936, and they felt that Vavilov was too conciliatory. A meeting with the director on November 11, held in preparation of the December genetics congress, instructed Vavilov to "give weighty words in a polemical manner."[33]

A general meeting of the scientific staff of VIR on November 19, 1936, demonstrated more divided opinions. Vavilov was not present and the meeting was chaired by the vice director.[34] His introduction praised the "initiative of the great scientist Lysenko" for socialist competition in science. This had dissipated the routine and torpor and raised the institute's work to a higher level. The main speaker was the geneticist Karpechenko, who discussed two of Lysenko's latest articles. He started by reminding the audience that VASKhNIL had found it necessary to conduct a broad discussion on the conflicting opinions about genetics in view of the international congress scheduled for August 1937. In these discussions, "a broad circle had been involved, frequently, regrettably, people incompetent in genetic questions."[35] Karpechenko concluded by saying that Lysenko was no doubt a great scientist who had made great contributions to agricultural science in the Soviet Union, but his genetics was "rubbish" (*erunda*).[36]

There were no objections or reservations to Karpechenko's positive evaluation of Lysenko's contributions to agricultural science. But Karpechenko's blunt dismissal of Lysenko's genetics was opposed by most of the speakers. Their main contention was that genetics had not been very helpful in furthering plant breeding and that the ideas of Lysenko deserved to be taken seriously. For instance, Evgeniia Sinskaia, a close collaborator of Vavilov, claimed that Lysenko's attack on genetics had its sound basis in a general "dissatisfaction with genetics among selectionists." Among the "sins" of genetics was that it did not consider the organism as a whole and its interactions with the environment. Genetics neglected the variation by "modifications," the variation brought on by different environmental influences. The "big fault of genetics was its distance from biology, from biological neighbor disciplines." That external factors could not produce the appearance of mutations, as Hermann Muller had recently claimed, seemed to Sinskaia "very strange . . . from a general dialectical point of view."[37]

Sinskaia complained that genetic literature was dull and uninteresting. Karpechenko replied that she had herself written the biggest and dreariest brick of a book ever published by VIR. Karpechenko's arrogant tone indicated tense scientific disagreements among the staff of VIR.[38]

One can speculate on the reasons for Vavilov's low profile in the genetics debate in 1935–1936. The young political activists within his own institute caused Vavilov great worry and took much of his energy. He had other reasons for muting his public criticism of Lysenko as well. Vavilov as well as some of his closest collaborators shared the view that classical genetics had fundamental weaknesses and needed radical reform to become a really useful instrument in plant breeding. Also, as late as the autumn of 1935, he had been strong in his praise of Lysenko, though not his genetics. Too sharp criticism now would appear as a contradiction of his own earlier claims. Vavilov was also a practically minded man who hated to waste time on speculative theoretical polemics that only cemented antagonism and animosity. A special reason for restraint on the side of Vavilov may have been fear of jeopardizing the international genetics conference. Close contact with international science would help the efforts to unveil the scientific weaknesses of Lysenko's ideas. If the internal Soviet conflict over genetics became too violent, the government might be tempted to cancel the congress. All these tactical considerations were certainly highly important for Vavilov's behavior, but disappointment with the contributions of classical genetics and dissatisfaction with the cocky behavior of some of its representatives also appears to have been an essential factor. Sinskaia probably reflected much of Vavilov's private thinking.

Futhermore, Vavilov as a non-Marxist and non–party member, had little authority in philosophical and scientific political debates. More important actors in the debate over genetics, and increasingly in discussions of science policy, were Communist Party members like Meister and Zhebrak, who really believed in applying dialectical materialism to genetics and plant breeding and Marxist doctrines in the politics of science. Their rather dogmatic defense of classical genetics and traditional principles in plant breeding was not in the spirit of Vavilov.

There was also another deep split among Lysenko's opponents. His geneticist critics ritually conceded the great practical services that he had given to Soviet agriculture, especially his vernalization of seed grain. But this method was a main point of attack for Lysenko critics among the agricultural scientists. The detailed and penetrating criticism by agricultural experts like Konstantinov and Lisitsyn tended to be neutralized by superficial praise from geneticists like Kol'tsov, Karpechenko, and Serebrovskii. Instead of facing Lysenko with hard facts about the vernalization of seed grain, the public discussions focused on methods of breeding and the general principles of genetics, areas with ample room for speculative hopes and wishful thinking about future progress. The theoretical and speculative character of the debates suited Lysenko's side and split the critics with internal disagreements. Among plant breeders there was widespread skepticism concerning the correctness and usefulness of classical genetics. Among biologists in general there was also widespread skepticism of classical genetics and sympathy for some of the fundamental ideas that Lysenko supported. We have seen that even a hard-core geneticist like Zhebrak was seriously striving to find a mechanism for the inheritance of acquired characters.

Thus there was no united front against Lysenko in 1936, simply because the critics were not able to find a shared scientific basis for it. Among leading biologists, few saw the full range of his scientific weaknesses. For instance, the eminent zoologist, cytologist, and experimental biologist Nikolai Kol'tsov could be surprisingly superficial with respect to plant science—perhaps he did not in general think much of botanists. In a 1937 fifteen-page general survey of the progress of Soviet biology during the preceding twenty years, Lysenko is the only plant physiologist, mentioned for his "brilliant experiments with the vernalization of winter plants." The vernalization of "winter varieties" on "enormous" fields was the practical result, according to Kol'tsov (1937a, 942). But as I have pointed out, it was the vernalization of spring grain and not winter grain that became Lysenko's big practical project. Even if this formulation was a slip of the pen by Kol'tsov, it does indicate that he did not pay much attention to the difference.

Most clear-sighted were perhaps practical breeders and agricultural experts such as Konstantinov and Lisitsyn. They attacked Lysenko where he was most vulnerable, demonstrating the false or unsubstantiated nature of his claims about the usefulness of vernalizing seed grain. These were factual questions which were understandable to a nonexpert audience and from which Lysenko could not so easily slip away with speculative theoretical arguments or ideological and political rhetoric. As seen from the outside, the lively debates within biology and agricultural research appeared complex.

190 | **THE LYSENKO EFFECT**

They presented no clear picture. No concerted struggle against Lysenkoism had yet emerged. To a nonscientist outside observer, Lysenko was likely to appear not as an extremist but as placed squarely in the middle of the scientific controversies.

NOTES

1. "Selection" was commonly used as a general term for the production of new varieties of cultivated plants. It covered the traditional methods of pedigree and mass selection as well as the use of hybridization and was more or less synonymous to the English term "plant breeding." In the following I will partly translate the Russian term *selektsiia* with "selection" and partly with "plant breeding."

2. See, for instance, Gregory (1917), Russian edition under the editorship of Vavilov (Gregory 1923).

3. "O selektsii i semenovodstve" (On selection and seed production), *Pravda*, August 3, 1931.

4. Letters from N. I. Vavilov to F. S. Stepanenko and T. D. Lysenko, March 29, 1932, VL 2, p. 165.

5. Letter from Vavilov to N. V. Kovalev, May 28, 1932, VL 2, pp. 173–74.

6. Letter from N. I. Vavilov to N. V. Kovalev, August 9, 1932, VL 2, pp. 179–80.

7. N. I. Vavilov, "Problema selektsii v SSSR," manuscript of lecture to be given at "Doma uchionykh" (House of scientists) in Leningrad on April 28, 1933. TsGAE, f. 8390, o.1, ed.khr 284, list 48ff.

8. Letter from N. I. Vavilov to "Kommissia sodeistviia uchionym pri SNK SSSR" (Commission on cooperation with scientists at the Cabinet of the Soviet Union), March 16, 1933, VL 2, p. 188.

9. See *Organizatsiia sovetskoi nauki v 1926–1939 gg, sbornik dokumentov* (Leningrad: Nauka, 1974), p. 364. The biologist was A. N. Severtsov and the chemist N. N. Semionov.

10. Letter from N. I. Vavilov to the biological section of the Academy of Sciences, February 8, 1934, VL 2, p. 219.

11. For instance, J. H. Woodger and E. S. Russell, see Roll-Hansen (1984).

12. Lysenko's general views on genetic theory can hardly be called

unreasonable for their time even if geneticists were unfavorable. He was in good company with many prominent biologists. It is significant that influential historical accounts of genetics still present the outcome of the controversy between biometricians and Mendelians as a compromise that was not clearly established until the neo-Darwinian "new synthesis" of the 1940s (Roll-Hansen 1989).

13. TsGANTD–St. Petersburg. f. 318, o. 1, d. 686. The year is given as 1934, but a more precise date is lacking.

14. TsGANTD–St. Petersburg, f .318, o. 1, d. 686, l.41–48.

15. This "meeting with the president" took place on July 17. A stenographic report is found in the VASKhNIL archive, TsGAE, f. 8390, o. 1, e. 604.

16. TsGAE, f. 8390, o. 1, e. 604, l.42–44.

17. TsGAE, f. 8390, o. 1, e. 604, l.58.

18. TsGAE, f. 8390, o. 1, e. 604, l.61–64.

19. TsGAE, f. 8390, o. 1, e. 604, l.77.

20. TsGAE, f. 8390, o. 1, e. 604, l.68.

21. TsGAE, f. 8390, o. 1, e. 604, l.82.

22. TsGAE, f. 8390, o. 1, e. 604, l.85–89.

23. TsGAE, f. 8390, o. 1, e. 604, l.89–93.

24. TsGAE, f. 8390, o. 1, e. 604, l.1–2.

25. *Izvestiia*, June 15, 1935, p. 2.

26. The declaration was addressed to Ia. Iakovlev, head of the agricultural department of the Central Comittee, M. A. Chernov, minister of agriculture, and A. I. Muralov, vice minister of agriculture and president of VASKhNIL. The declaration was signed by four people: T. D. Lysenko, as "scientific leader of the Institute for Genetics and Selection," F. S. Stepanenko, as "director of the instititute," one representative for the local party organization, and one for the trade union. (*Iarovizatsiia* 1 [July–August 1935]: 3–5.)

27. The article referred to was Savchenko-Bel'skii, "Perestroika osnov agronauki," SZ 133 (1935). (Meister 1935a, 16.)

28. See chapter 5.

29. TsGAE, f. 8390, o. 1, e. 604, l.94.

30. Coauthored with Lisitsyn and Kostov.

31. An example was P. N. Iakovlev, author of an article "On the Theories of 'True' Geneticists," immediately following Vavilov's in the same issue of *Socialist Reconstruction of Agriculture*.

32. TsGAE, f. 8390, o. 1, e. 955. l.34–35.

33. Popovskii (1983, 113). Popovskii quotes a protocol from a meeting with the director of VIR on November 11, 1936.

34. A. B. Aleksandrov was vice director in 1935–1937, recommended by Iakovlev.
35. TsGANTD–St. Petersburg, f. 318, o. 1–1, d. 1133, l.2–3.
36. TsGANTD–St. Petersburg, f. 318, o. 1–1, d. 1133, l.14b.
37. TsGANTD–St. Petersburg, f. 318, o. 1–1, d. 1133, l.20–23.
38. TsGANTD–St. Petersburg, f. 318, o. 1–1, d.1133, l.72–79.

Chapter 7.

"Two Directions in Genetics" — The Congress of December 1936

The controversy over genetics took a new turn at the end of 1936. The onset of the great political terror of 1936–1937 politicized and polarized debates both on the science policy and on the scientific levels. At the VASKhNIL congress of December 19–27, a pattern of two opposed camps emerged more clearly than before. Before this congress the discussions within VASKhNIL had been broadly concerned with practical questions, even if there was the tendency toward theoretical, methodological, and ideological generalizations that I have described in preceding chapters. Ironically, the strong ideological insistence on "unity of theory and practice" had the effect of turning attention toward theoretical problems of method and organization and away from the practical problems of agriculture. At the end of 1936 the debates in VASKhNIL became focused on questions of genetics. Sensitive questions of human genetics and eugenics were drawn in[1] and the debate became polarized between "two directions in genetics." Nikolai Kol'tsov and Hermann Muller entered as central participants. This turn toward genetics pertained in particular to the open public debates, and in particular it characterized these debates as they were reported by the mass media and perceived by the nonexpert public. In the politically tense situation, the repercussion of media presentations on the internal scientific and policy debates of VASKhNIL was strong.

Since early 1935 Lysenko and Prezent had pursued a line of confrontation with their proposals for new breeding methods and increasingly radical criticism of classical genetics. On the eve of the great terror of 1936–1937 the already limited space for objective and open scientific debate became even narrower than it had been since the great break of the late 1920s. Arguments that appealed to partyness and simplistic practicality gained in influence. The one-sided picture of classical genetics painted by its adversaries under the label "formal genetics" obtained a kind of offical status (Muralov 1936c, 2). The term had been used earlier by geneticists themselves to designate views and tendencies that they did not accept.[2] Now the negative connotations covered the whole of classical genetics. Vavilov was reluctantly being pushed into a position of leading the defense of classical genetics, which he viewed with considerable reservation. As we have seen, his reserved attitude was typical of many leading plant breeders both in the Soviet Union and in the West. Vavilov was not in doubt about the fundamental theoretical achievements of classical genetics and believed strongly that it would give abundant practical benefits in the long run. But he had warned against excessive short-term expectations and stressed the need for searching theoretical criticism and revision. He had tried to hold back enthusiasts like Meister and Karpechenko.

Just ahead of the congress, VASKhNIL president Muralov made it clear that the common platform of debate was to be "the Marxist-Leninist-Stalinist worldview," including dialectical materialism and the rejection of "fascist 'theories' of race." On this basis, a comprehensive examination and evaluation of different genetic theories was to "provide unity of method" for practical breeding work (Muralov 1936c, 2). Muralov's formulation of the platform for the conference revealed an unrealistic belief in the ability of science to decide pending controversies simply by a comprehensive examination of all known facts and theories. He was affected by a superstitious belief in scientific rationality similar to that of Lysenko at the same time as his knowledge of genetic theory and facts was even more limited. Muralov's presentation included a confused description of classical genetics as holding that the gene "can undergo change only in the case of the uneven division of cells and disturbance of the distribution of chromosomes (by grafting)" (Muralov 1936c, 2).

The congress was dedicated to "questions of breeding and genetics," which were to be posed not in an "abstract or scholastic fashion," but concretely linked to the practical breeding work. It was to "survey the most important results in the field of selection," and the methods that science provides to breeding work were to be evaluated from the point of view of their practical efficiency. This was Muralov's message in his opening address. He reminded the audience that at the Saratov meeting in October of 1935 VASKhNIL had given to the geneticists "the task to help make new varieties of grain and high-yielding cattle as rapidly as possible." Since then leading institutes for plant and animal breeding had been "subjected to the strongest criticism" and were expected to respond. Discussion of the principles of genetics had a special significance because this science had so often been subject to "stagnation and conservatism," said Muralov. Once again he quoted Stalin's words to the Stakhanov workers in November 1935: "Science is called science because it does not recognize fetishes, does not fear to lift its fist against the obsolete, and closely listens to the voice of experience, practice"[3] (Muralov 1936d).[3]

The political importance of the congress was marked by the participation of officials at the opening: Iakovlev, the party chief of agriculture; Chernov, the minister of agriculture; Bauman, the party chief of science; and his deputy, Doroshev. A new scientific secretary of VASKhNIL, L. S. Margolin, had taken over from Bondarenko. The tightening political situation was signaled on the opening day of the congress by the news that Israil Agol, a young communist geneticist and science administrator, had been arrested.[4]

The first half of the congress consisted of reports on practical results in plant breeding. The debate on the controversial issues of genetics came in the second half. This part started with four main lectures by Vavilov, Lysenko, Serebrovskii, and Hermann Muller, in this order, followed by three days of general discussion.

The American geneticist Hermann Muller backed the geneticists' cause with his international scientific authority. He had been a core member of Thomas H. Morgan's group that established the chromosome theory. In 1927 he published his demonstration of artificial mutations, for which he received the Nobel Prize in 1946. Socialist political sympathies, plus academic and personal difficulties, brought him to the Soviet Union. From 1933 he worked in the Academy of Sciences' Institute of Genetics headed by Vavilov.[5] Muller was also an ardent eugenicist who hoped that the Soviet Union was a place where a truly humanistic and socialist eugenics could be realized. In May 1936 he sent a copy of his eugenic tract *Out of the Night* with an accompanying letter to Stalin asking him to consider a program of artificial eugenic insemination.[6]

The congress officially had seven hundred participants, but the second half was moved to a larger auditorium because of the great public interest, and here the audience was reported to be more than three thousand.[7] *Pravda*, the Communist Party newspaper, gave a broad coverage, printing daily articles about the proceedings with a clear preference for reporting the views of the Lysenkoist camp. *Izvestia*, the paper of the government (council of ministers), published only short versions of the lectures of the two main figures, Lysenko and Vavilov.

From the viewpoint of the political and administrative leadership of VASKhNIL, a main purpose of the congress was to sum up results from the socialist competition in plant breeding. They hoped the practice criterion of truth could help sort out the conflict over plant breeding that was threatening to paralyze VASKhNIL. Muralov's statements showed a clear preference for Lysenko's standpoint and revealed a high degree of ignorance in genetic science. The wording of his misunderstandings also indicates a lacking realism about the limits of his own scientific insight. Still, the political leadership and bureaucracy depended on some kind of agreement among the scientists. As long as there was open conflict between large factions of the scientific expertise, they did not have the authority to enforce a solution.

LYSENKO VERSUS VAVILOV

In his lecture on the fourth day of the congress, Vavilov concentrated on describing and defending the work of VIR. His discussion of Lysenko's ideas on genetics and breeding played down the controversial issues, though he did not completely overlook them. Partly, he said, the disagreement was caused by differing definitions, for instance, of "inbreeding." For Vavilov, "inbreeding" did not have to be harmful. It could be a very useful method. He also defended Johannsen's theory of the stability of pure lines and maintained that selected varieties, which approximate pure lines, were not subject to the kind of degeneration that Lysenko claimed. There were no data that contradicted traditional methods on this point (Vavilov 1937, 39-41). As long as Lysenko had not presented precise and substantial experimental results, it would thus be wrong to apply his method of artificial intravarietal cross-fertilization.

Part of Vavilov's strategy was apparently to win time. According to Muralov, the correct resolution of the disputed issues had "enormous" importance, not only in theory but also for practice (Muralov 1936c, 2). Vavilov tried to neutralize this impression of urgency by saying that there was sufficient agreement for the practical work to continue, without any immediate large changes. Partly the controversy was nothing more than a normal disagreement between scientists, which could not properly be decided by a congress because more experiments and other scientific works were needed. Partly there was sufficient agreement for immediate practical action. In this way Vavilov tried to evaded just those issues that the congress had officially been called to discuss.

> Apart from disputed issues in those areas where there are not yet sufficient experiments, much is completely undisputed. In essence the road by which to proceed in the development of selection work is fundamentally clear to Soviet selectionists. (Vavilov 1937, 42)

But in the tense political atmosphere, Vavilov's restraint was not well received. Among geneticists and Lysenkoists, as well as politicians and administrators, the general opinion was that Vavilov's lecture was too weak and evasive. He had disregarded the main purpose of the congress.

Lysenko, on the other hand, was uncompromising in his lecture "On Two Directions in Genetics." He posed as the practically oriented scientist who had the masses behind him and was generously applauded for his gibes at academic science.[8] Lysenko insisted that the controversy concerned fundamental scientific attitudes and principles with important ideological as well as practical consequences. His lecture was a continuous polemic against Vavilov. In spite of the moderation he had shown, Vavilov was staged as the main opponent. He was the man who had "fought against" Lysenko's results in the breeding of spring wheats, that is, the new varieties he claimed had now been amply confirmed. "It is incomprehensible to me," said Lysenko, "that Vavilov insists on his false views even after decisive practical proof. Now they are not only false, they are harmful."[9]

It was the very theory of evolution that was in dispute, according to Lysenko. In capitalist societies Darwin's theory of the origin of species had been under constant attack from the time it was published. The critics seized upon minor inaccuracies or defects and then attempted to pervert the whole theory by "correcting" it. As an example of such perversion, Lysenko pointed to Johannsen's theory of pure lines and claimed that it conflicted fundamentally with Darwin's view of the variability of species.

Lysenko claimed that the evidence for genetic constancy in Johannsen's classical selection experiment with beans was quite insufficient. He sketched an alternative experiment that would certainly have prevented Johannsen from drawing his mistaken conclusion:

> If Johannsen had propagated his selected lines of beans to more substantial quantities, perhaps to a zentner, and then applied selection of the extreme variants, taking into consideration the conditions of development for the selected plants, or if he merely had applied selection on a considerably larger scale, then he would under no circumstances have arrived at a conclusion telling that there is no value in selecting among self-fertilizing plants which descended from one seed. (Lysenko 1937, 35)

In other words, the scale of Johannsen's experiment was too small. Lysenko simply asserted this as a thought experiment based on his own assumptions, among them the principle that large-scale, practical experience is superior to purely scientific experiments in deciding the truth of theories.

The fundamental mistake of the geneticists, said Lysenko, was that they denied a "creative role for selection in the evolutionary process." This was denied not only by Johannsen but, for instance, also by the American geneticist T. H. Morgan, and the view was shared by Vavilov and his associates. On this issue, said Lysenko, he and those who agreed with him took the stand of Darwin's theory of evolution and were in fundamental disagreement with the school of Vavilov. In contrast to Vavilov, Lysenko's tone was uncompromising:

> On this question there is, between these two directions in science, a principal difference which it is impossible to reconcile through agreement on particular isolated and minor questions. (Lysenko 1937, 36–37)

For Lysenko the creative role of selection was the touchstone of Darwinian orthodoxy. The difference in views on this point could not easily be bridged. From this fundamental theoretical difference derived the other points of conflict, according to Lysenko. His insistence on natural selection as a "creative force" echoed the debates in the West, where it has also been common up to the present day for more-orthodox Darwinists to accuse geneticists of denying or neglecting the "creative force" of natural selection. No doubt Prezent's teachings of "creative Darwinism" was a source of inspiration for Lysenko's arguments.

Johannsen's theory of pure lines was attacked because it was the basis of traditional seed growing and selection as well as a direct contradiction of Lysenko's claims about the degeneration of self-fertilizing varieties. Johannsen's more general theory about the relationship between genotype and phenotype was also incompatible with Lysenko's ideas on how to effect a directed change of heredity.

In his lecture Lysenko discussed two concrete problems: "improvement of the quality of the seed-material of self-fertilizers through hybridization within the variety" and "alteration of the nature of plants in the direction we need through a suitable education." He was no lover of theoretical discussions, he said, and therefore limited himself to those cases where such discussions are necessary to solve the practical tasks one has set oneself. On the latter problem, Lysenko now challenged the geneticists with an experiment with direct theoretical importance. He claimed to have shown how heredity could be changed in a chosen direction by suitable manipulation of the environment. The line of thought in his experiment was quite similar to the ideas about modifications gradually becoming hereditary that had been sketched by Boris Keller and Anton Zhebrak.

Lysenko believed that during the vernalization stage, when the crucial influence of low temperature took place, the hereditary basis for this process would also be malleable to some extent. Vernalization takes place in the cells at the growing point, explained Lysenko. These vernalized cells then give rise to the parts involved in reproduction, ending up with the ripe seeds. If two plants are subjected to different temperature regimes during the process, this will produce differences in their vernalized state, and this difference will be transferred to all descending cells in some form. These differences will also be reflected in the germ cells and in the seeds. According to Prezent, continued Lysenko, the environmental requirements of a plant to some extent repeat and reflect the developmental path of its predecessors. And the closer a predecessor is, the larger its influence. Thus, reasoned Lysenko, if one takes a winter wheat, which for thousands of generations has been vernalized at relatively low temperatures and subjects it to vernalization at higher temperatures, its heredity will change in the direction of a spring wheat (Lysenko 1937, 53).

The experiment to change the winter wheat Kooperatorka into a spring wheat had been started on March 3, 1935, Lysenko now revealed. Two plants of Kooperatorka, together with two plants of another more extreme winter variety, Liutestsens 329, had been planted in a flower pot and placed in a nonheated greenhouse, where the temperature did not rise above 10 to 15 degrees centigrade until the end of April. While the plants of Liutestsens 329 died in the late autumn without producing straw and spikes, one of the two Kooperatorka plants gave ripe seed by early September. (The other plant perished because of "pests gnawing its roots.") These seeds were planted on September 9 in a warm greenhouse. The plants developed more spikes than controls of ordinary Kooperatorka, but more of the spikes were sterile and the awns were shorter. Already at this sowing, it was clearly seen that "the experimental plants were different in their behavior, in their nature, from the controls." On March 28, 1936, the second-generation offspring of the original Kooperatorka, called "third generation," was sown together with the offspring of the controls, called "second generation," and ordinary Kooperatorka as control. The result of this last step in the experiment was that the "third generation" developed more quickly and fully than the "second generation," and the controls died in the autumn without producing any spikes. The "third generation" gave a normal development of all shoots, while the "second generation" was much retarded. The behavior of a "fourth generation" sown in September 1936 confirmed that the Kooperatorka had undergone a clear change in the direction of a spring variety. Lysenko also reported in passing that the alteration of winter rye into spring rye proceeds more quickly and easily than that of wheat (Lysenko 1937, 54–58).

The limited material alone was enough to disqualify this as a serious scientific experiment. With only one plant surviving the crucial stage, the evidence was weak, indeed. The possibility of accidental mistakes was obvious, and the need to repeat the experiment before any conclusions worth further attention could be drawn should be quite clear according to normal rules of scientific method. As we now know, Lysenko and his followers were unable to repeat the experiment in any convincing way. Lysenko nevertheless concluded his lecture by dogmatically rejecting the chromosome theory and pronouncing a holistic view of inheritance: "The hereditary basis is the cell which develops and is transformed into an organism." With the microscope it will be possible "to see more and more details in the cell, in the nucleus and in the chromosomes, but these will be pieces of the cell, nucleus and chromosomes, and not what the geneticists understand by a gene," argued Lysenko (1937, 66).

COUNTERATTACK OF THE GENETICISTS— ENTER KOL'TSOV AND MULLER

In their talks, Serebrovskii and Muller[10] gave expositions of classical genetic theory and the experiments it was built on. They both claimed there was very little doubt that the genes existed as structured particles, a kind of macromolecules, distributed along the chromosomes. And they both explained how mutations in the genes provided the fundamental variation that was necessary for evolution to proceed. The relative stability of the genes was not in contradiction to the phenomenon of biological evolution. Serebrovskii ridiculed the superficial and ignorant view that Lysenkoists took of hereditary variation. They claim that hereditary change can be produced by "grafting, feeding, vernalization, and whatnot of deformations" and that these changes have precisely the adaptive character that would be so convenient for our breeding work. And then they ask: Why do the geneticists not use these methods rather than break their heads over chromosomes, genes, x-rays, and radium? The answer, said Serebrovskii, is simply that the methods that the Lysenkoists recommend have already been tried with a negative outcome. Tens of scientists had spent tens of years on this task. Tens of times it had seemed as if they had obtained the results they hoped for. And every time there had been bitter disappointment as they discovered their mistake, said Serebrovskii. In other words, the Lysenkoists misjudged the situation because they did not know the scientific tradition. Bluntly put, they were scientific illiterates.

Muller did not deliver his lecture personally. It was presented by Nikolai Kol'tsov, who was the grand old man in Russian experimental biology. Kol'ltsov had worked primarily with the new experimental and microscopic investigations of the cell developed around the turn of the nineteenth century and the new general biological theories linked to them. To the extent that he had experience in practical science, it was in husbandry and medicine. In contrast to Nikolai Vavilov, he was a zoologist grounded in the basic theoretical questions of biology. He was also fifteen years older than Vavilov, with a career as a radical young scientist and supporter of the failed revolution of 1905. As a brilliant researcher, teacher, and scientific entrepreneur, he had formed the development of experimental biology in the Soviet Union of the 1920s and 1930s more than anybody else.[11] Like many other biologists at the beginning of the twentieth century, Kol'tsov had been fascinated by eugenics. During the 1920s his main source of economic support was the Ministry of Health. The possible applications of genetics to medicine and social policy was an important political motivation for this support. When eugenics became linked to fascism at the beginning of the 1930s, Kol'tsov quit the field. But in the ideologically highly charged atmosphere of the Soviet Union in the 1930s, past interests in eugenics became an effective argument against him and against classical genetics in general.

Muller's lecture was a brilliant but rather dogmatic popularization of the principles of classical genetics. Clearly, many biologists and breeders, and not only the followers of Lysenko, found his treatment of the debated questions, such as the stability of the gene and the inheritance of modifications, highly arrogant. After Kol'tsov had presented the formal lecture, Muller himself gave some final comments characterizing Lysenko's ideas as "quackery," "astrology," and "alchemy." Muller's main provocation, however, was to accuse the Lysenkoists of unacceptable views on human genetics. The congress had been instructed not to discuss questions of eugenics and human genetics (Adams 1990b, 196). But once the instruction had been violated by Muller and Kol'tsov, the Lysenkoists exploited the topic eagerly in the ensuing debate.

A few months later Muller described in a letter how he had "called attention to the fascist race and class implications of Lamarckism, since if true it would imply the genetic inferiority, at present, of peoples and classes that had lived under conditions giving less opportunity for mental and physical development." The audience "applauded wildly, but there was a terrific storm higher up and I was forced to make a public apology, while the statement was omitted from the published address." And Muller did not stop at this. When Serebrovskii later in the conference was forced to repudiate his 1929 program for a socialist eugenics, Muller stood up and defended some of Serebrovskii's earlier views. But the translator had slurred over what he said, and he was informed that his statement could not be printed in the proceedings, explained Muller to his friend the British zoologist Julian Huxley.[12]

Muller "played directly into the hands of the Lysenkoists," according to the American historian of science Mark Adams. By resurrecting the argument that Iurii Filipchenko and Serebrovskii had used against Lamarckism in the 1920s, Muller "helped reestablish the ideological links between eugenics and genetics," and he compromised some of the leading geneticists by drawing attention to their earlier eugenic interests. This applied to "his senior colleague Kol'tsov, his patron Vavilov, his friend Serebrovskii, and his student Levit" (Adams 1990b, 197). Human genetics and eugenics had already become a central issue in the struggle over the coming Seventh International Congress of Genetics, scheduled for Moscow in August 1937.

In the general discussion following the main lectures, Kol'tsov defended theoretical genetics. It is common to speak about selection *or* genetics, theory *or* practice, but this is misleading, he said. Both are necessary. The purely theoretical genetic investigations of the fruit fly, *Drosophila*, are important because they provide the quickest and cheapest access to the basic laws of heredity. Both as teachers in universities and as advisors in practical affairs, scientists trained in such theoretical studies are very valuable, explained Kol'tsov. He illustrated his argument with two telling examples.

Even the most brilliant practical selectionist cannot fulfill the tasks that the Soviet government sets without knowledge of recent scientific achievements, including those of bourgeois countries, said Kol'tsov. Coming to this conference, he had wondered if anybody would refer to the results from the congress of potato specialists that had taken place in England three months earlier. England had similar problems as the Soviet Union with the degeneration of potatoes. Four different viruses had been discovered, and a particular species of plant louse had been identified as the vector for three of them. Kol'tsov had not found anybody among the present potato specialists who knew about these results. Perhaps vernalization and summer planting has a positive effect because the virus is usually transferred in the spring, and by summer it is too late for the lice to infect the plants, speculated Kol'tsov. Well, we do not know, this has to be investigated (Kol'tsov 1937b, 239–40). The critical point and the address of this example were clear. Vernalization and summer planting of potatoes had been one of Lysenko's proposals for the last couple of years.

Kol'tsov's second example was an anecdote about the world-famous and recently deceased physiologist Ivan Pavlov and how he learned about genetics. Genetics is a young science, and Pavlov was an old man. He had not learned much as a student, and his research had been in other areas. It was not surprising that his insight was superficial. Now one of Pavlov's students found that the offspring of mice that had learned to react to a bell learned more quickly than their parents. In spite of Kol'tsov's warnings, Pavlov had presented this abroad as a case of Lamarckian inheritance. But he was corrected by the American geneticist T. H. Morgan. Kol'tsov also wrote an article explaining the mistake. It turned out not to be a case of the mice becoming able to learn more quickly but of the experimenter gradually learning how to teach the mice more effectively. Control experiments with mice descending from parents who had not been taught showed them to learn just as quickly. Pavlov, however, had smilingly received Kol'tsov in his laboratory and admitted the mistake. "You were right, I know dogs and will forever work with them, but with mice I will not work." Pavlov was a humble scientist who knew that he could make mistakes and had the courage to admit them (Kol'tsov 1937b, 240–42).[13] The obvious contrast was Lysenko.

To make the message clear, and perhaps to remove some of the sting for Lysenko, Kol'tsov turned to Vavilov and asked: Nikolai Ivanovich, "do you know all the genetics that is needed?" No, "if you were given an ordinary student excercise to determine the location on the chromosome of a particular mutation, you would, excuse me, not be able immediately to solve this problem, since there were no student courses in genetics in your time" (Kol'tsov 1937b, 243).[14]

The cytogenticist Grigorii Levitskii, on the staff of Vavilov's VIR, defended the concept of material corpuscular genes in the form of large molecules. He told about the recent discovery of the giant chromosomes in the salivary glands of fruit flies (*Drosophila*). Here one could see in a microscope the specific morphological structures that corresponded to the genes, he argued. Most of his contribution was a polemic against Prezent's attacks on the concept of a particulate and material gene (Levitskii 1937).

SHADES OF SYMPATHY FOR LYSENKO

At the December 1936 conference, there was not yet a sharp division into two camps, for and against Lysenko, even if Lysenko and his followers tried to create this impression. There was considerable agreement with some of Lysenko's criticisms of classical genetics among prominent biologists who were quite independent of Lysenko and cannot be said to belong to his followers. Most of them later turned against him.

Among those who went furthest in supporting Lysenko was the embryologist Boris M. Zavadovskii. He had defended the inheritance of acquired characters in the 1920s, and now he explicitly took a position against classical genetics and in favor of Lysenko's views. Zavadovskii emphasized the insufficiency of the mechanistic type of explanation in classical genetics. He found Lysenko's research program to be a good starting point for a highly needed reform of genetic theory as well as a basis for practical breeding work. For instance, he defended Lysenko's proposal for cross-fertilization within the varieties of self-fertilizing cereals (Zavadovskii, B. M. 1937).

Boris Zavadovskii was the brother of Mikhail Zavadovskii, the VASKhNIL vice president. The latter continued his sharp criticisms at the December conference. Once more M. Zavadovskii put the blame on Prezent for leading Lysenko astray and approached Lysenko with conciliatory comments. Some comrades demanded a choice: either Vavilov or Lysenko, said Zavadovskii. But his thesis was *both* Vavilov and Lysenko—we need Lysenko's enthusiasm and talent. Zavadovskii concluded by claiming that a synthesis between genetics and the theory of individual development was needed to correct the inadequacies of classical genetics. Like a sleepwalker Lysenko was heading in the right direction (Zavadovskii, M. 1937). Apparently, M. Zavadovskii's attitude and strategy at this point was much like Vavilov's.

Vavilov's coworker E. N. Sinskaia defended VIR while taking a jaundiced view of theoretical genetics. She welcomed Prezent's attempts to unify phylogenesis, the theory of descent, with genetics, the theory of inheritance. Sinskaia thought it quite possible that the stages of individual development contained the key to directed change of heredity. Lysenko's claims on this point had to be taken seriously and properly investigated. The outright rejection of Lysenko by the geneticists was not well founded (Sinskaia 1937).

The plant physiologist and morphologist Nikolai P. Krenke was also quite skeptical of classical genetics, without showing much confidence in Lysenko's ideas. Quoting extensively from T. H. Morgan's recent book on embryology, Krenke argued, like the brothers Zavadovskii, that there was indeed much more doubt about the theoretical foundations of genetics than appeared from the lectures of Serebrovskii and Muller. The two had given a quite misleading picture, for instance, with respect to the stability of the gene (Krenke 1937, 303). However, the fact that the variability of the gene was an open question was no reason to reject "formal genetics," according to Krenke. First, the knowledge that classical genetics had produced would have to be explained and reproduced on the basis of an alternative theory (304), and this had not been achieved by Lysenko. Krenke pressed Lysenko hard to give a clear explanation of his claims concerning the numerical relationship in segregating hybrids, in particular the famous Mendelian 1:3 relationship that Lysenko often claimed to have disproved. You say that your claims have been "incorrectly understood," said Krenke. "Well, we do not understand you, neither Levitskii, Karpechenko, nor I, and many others." But "we are no idiots, and you should try to explain so that we understand what you are saying" (Krenke 1937, 308).

Krenke ended his contribution by defending the freedom of scientific expression. He illustrated its practical importance through an example from his own experience. At a conference that discussed procedures for cultivating certain rubber-carrying plants, he alone had maintained that the proposed procedures were biologically impossible. Because of this obstructive behavior, he had become unpopular among administrators and researchers in the field. But a few years later the director of the program had called on him to help with a fundamental revision of the plans that had been based on the outcome of the conference. Some of them had turned out to be "objectively harmful." Krenke urged sincere and open expression of scientific views. Sincerity and perseverance, in spite of difficult conditions, is "the best path for our socialist development" (Krenke 1937, 310–11).

The old generation of plant breeders was generally negative toward Lysenko's new ideas. But the younger ones were more positive. Nikolai V. Tsitsin was born in 1898, like Lysenko, and had acquired scientific fame and prizes for his work on interspecies hybrids. His hybrids between wheat and couch grass attracted great attention as he developed them into perennial varieties of wheat—which in the end turned out to be a disappointment in practical agriculture. This failure did not detract from Tsitsin's achievements as a pioneer in the investigation of interspecies hybrids, but it probably sobered his enthusiasm for Lysenkoism. He eventually left plant breeding and became director of the Moscow botanical garden, retaining considerable influence in Soviet biology.

At the December 1936 conference, Tsitsin claimed that he had obtained results that confirmed Lysenko's ideas about the usefulness of hybridization within varieties. He also gave some theoretical arguments why similar positive results were likely in many other cases. Furthermore, seed production in the Soviet Union was in a bad state despite many attempts at improvement. Lysenko's ideas seemed a good starting point for much-needed reforms. Between the two directions that existed in genetics and selection, the choice was obvious for Tsitsin. Only one of them respected the truths that all science depends on social class and that every scientific theory is proved by practice. Serebrovskii had spoken with pathos about genes, but they remained a thing apart from practice, said Tsitsin. Promises to solve the problems of socialist agriculture some time in the future, perhaps in fifty years, were of little help (Tsitsin 1937).

SUMMING UP: MEISTER'S COMPROMISE AND MURALOV'S "BASIC RESULTS"

In a concluding contribution to the general discussion, Vavilov was considerably clearer in his criticism of Lysenko than he had been in his main lecture. He listed and explained a series of points of disagreement. But again he ended on a conciliatory note. The differences have become clearer, and what is now needed is more mutual respect and attention to each other's work, he said: "Though we disagree on some theoretical questions, we have the same aim: we want in the shortest possible time to change cultivated plants" (Vavilov 1937, 49).

The conclusions that Lysenko pressed in his concluding statement were of a different nature and tone. To him the discussion had revealed that the leading geneticists did not know their classics, Darwin and Timiriazev in particular. Genetics contradicted the evolutionary theory of Darwin and was therefore unacceptable (Lysenko 1937, 66, 75). He did not attempt any detailed debate with his critics.

VASKhNIL vice president and genetics expert Georgii Meister was given the task of summing up the results of the conference. After the general discussion, he presented an evaluation of the various contributions, handing out praise and criticism. His lecture was officially approved by Muralov and printed in a number of newspapers and journals. Meister saw that Muralov's demands for clear answers on the debated issues were unrealistic and ended up balancing the two sides. Each side had its strong and weak points. Meister behaved much like another party member and leading plant geneticist, Zhebrak, emphasizing the theoretical and methodological issues, and still not forgetting the more detailed scientific issues and the empirical evidence. The emphasis on philosophical and ideological issues distinguished their approach from that of Vavilov.

Meister sided on some issues with Vavilov and on others with Lysenko—roughly, with Vavilov on concrete and substantial scientific issues and with Lysenko on a number of methodological and science policy issues. Meister said that he would not try "to reconcile principally different views" but rather to bring out "the healthy core" in each of them. He wanted to stimulate an objective scientific debate without attempts to disarm the opponent by the use of "isms." But the heated nature of the debate worried him. Both in the introduction and in the conclusion of his lecture, he warned against personal attacks and tendentious interpretations of opponents (Meister 1937, 4).

Like Lysenko, Meister appealed to Darwinism. But he also reminded the audience that Darwin was bound to represent the state of knowledge in his own time and that he naturally had been mistaken on many points. At the Odessa institute there was a tendency to take Darwin too literally, for instance, with respect to issues of self-fertilization, said Meister (1937, 5).

Meister praised Serebrovskii's and Muller's defense of genetics. They were excellent in many ways. But they had a tendency to forget the broad perspectives of Darwinism, the "historical point of view." It was very unfortunate for the ongoing discussion that they had focused so much on the stability of the gene that it became difficult to understand how heredity could change. Muller should have concentrated on explaining that the change in the genes, the mutations, are random and not adapted to the environment. Furthermore, in Meister's view, both Serebrovskii and Muller had neglected the role of the environment in evolution. As Darwinists we have to recognize "the creative role of selection," said Meister (1937, 6–7), echoing the insistence of Prezent and Lysenko that natural selection is "a creative force."

Though Serebrovskii had retracted his eugenic errors, Meister found it appropriate to discuss them. Meister saw a link between Serebrovskii's neglect of environmental factors in evolutionary theory and his tendency toward a simplistic biological thinking about human society. Meister emphasized that biology should serve man, and that Marxism could not accept its pretensions to a leading role in social questions. Serebrovskii's application of principles of animal breeding to man by proposing large-scale insemination of women with the sperm of outstanding men, was "deeply insulting to Soviet women," however genetically clever it might be. If the memory of this false step survived into history, Serebrovskii himself was responsible, said Meister (1937, 8).

Meister also defended Lysenko against Muller's accusation that his Lamarckian theory of inheritance supported the Nazi ideas about inferior races. Any theory can be used for such evil purposes as the Nazis have in mind, argued Meister. The theory of pure lines and stable genes fit the Nazi doctrines even better than that of Lysenko. In any case Lysenko was a loyal Soviet citizen, and the attempt to connect his theory with Nazi ideology was pure slander (Meister 1937, 14).

Vavilov's introductory lecture was highly unsatisfactory in Meister's view. It evaded the problems that the congress had been called to discuss and instead presented a quite unnecessary defense of VIR. There was general agreement that VIR was a useful and valuable institution, especially its World Collection. The question was how the work of the institution could be made as efficient as possible—Meister presumably had the destructive internal conflicts of VIR in mind. And Vavilov had completely avoided discussing future plans for the institute. He had not even subjected his own theories to a critical examination. In brief, Vavilov's lecture was generally lacking in self-criticism, concluded Meister (1937, 10–11).

The methodological mistakes of Vavilov were important because he was such an influential person, said Meister. He criticized in particular the law of homologous series, claiming that it was built on "atomistic conceptions" and "a theory of invariability of the gene." The anti-Darwinian tendency of the law worried him. If Vavilov had only based his work fully on materialistic Darwinism, "we would have been spared hearing that the law of homologous series supports the idealist views of Berg" (Meister 1937, 12). (Lev S. Berg was an eminent biologist critical of natural selection as the main explanation of evolution.)

Lysenko was praised by Meister for his work in plant physiology. In particular, vernalization and the theory of stages had been important for selection. Lysenko's general efforts to link "theory with practice" were also exemplary. But it is significant that Meister was highly critical with respect to vernalization as a profitable agricultural technique. Konstantinov's careful criticism of the vernalization of seed grain, which was repeated at the conference, received the unreserved support of Meister. He also pointed to the unscientific nature of Lysenkoist attacks on Konstantinov (17–18).

Within breeding and genetics, the areas where Meister was an expert, his judgment of Lysenko was generally negative. Lysenko's critical attitude toward traditional methods of plant breeding was useful (Meister 1936, 17). But he sharply rejected the specific claims. Lysenko lacked evidence that pure lines degenerate or that self-fertilization is detrimental. "We cannot permit a groundless condemnation of theories and methods whose practical value have been established through broad scientific experience," warned Meister. Meister was also worried about the freedom of the press. "We know that some not so well-considered statements by Lysenko on pure lines, inbreeding, and genetics have in certain places been interpreted to mean that work with inbreeding is almost a counterrevolutionary activity, and some of our newspapers and publishers refuse to print articles about inbreeding and even about genetics" (Meister 1936, 15).

Meister met Lysenko's alteration of winter to spring wheat with an open mind. He did not believe it was a case of the inheritance of modifications, as Lysenko claimed. But we all know Lysenko as "a scientist with an outstanding power of observation," said Meister, and he would not exclude that "with this experiment new ways of changing heredity have been discovered." He made it clear, however, that the experiment had to be "verified with the participation of physiologists and cytologists" (Meister 1936, 17).

In his conclusion Meister emphasized that the controversial issues had not been resolved. Future progress demanded scrupulous empirical investigations by competent researchers.

> I believe that on a number of questions our discussion is far from ended. But I am convinced that it will be continued with detailed investigations of experimental results and the testing of disputed issues by sufficiently competent workers. (Meister 1936, 19)

Meister, like Vailov, trusted that future investigations would confirm that classical genetics was basically right and Lysenko's ideas on heredity generally mistaken. Even if it proved impossible to convince Lysenko, the political authorities would respect the scientific results. But unlike Vavilov, he had a strong belief that these investigations could best proceed within a conceptual framework of Marxist theories about science and under the political guidance of the Communist Party.

However, it was a politician-administrator and not a scientist who had the last word at the conference. In a short final speech, "The Basic Results of the 4th Session," Muralov summed up. Again the vague and superficial character of the speech's scientific judgments reflected Muralov's lack of scientific understanding. Lysenko's new methods of plant breeding was given the place of honor. The Odessa institute was exemplary for its development of methods "for speeding up the breeding of new varieties, for their rapid propagation and introduction into production" (Muralov 1937, 3). Muralov's statement of the "main results" was an act of political balancing, rather than the scientific judgment it pretended to be.

Lysenko's main claim to new methods in plant breeding was the new wheat varieties that he had demonstrated at Odessa in July 1935. After sharp criticism by Meister and others following this demonstration, Lysenko had been silent on that specific claim. The varieties from 1935 were not mentioned in Lysenko's lecture at the December 1936 session. Instead, he launched new speculative methods and "results" without substantial empirical evidence. But so far Muralov apparently took little notice of Lysenko's jumping from one project to another without pursuing any of them to a successful completion.

POLITICS BEHIND THE SCENES

The ideological bosses of the party Central Committee did not take part in the public scientific discussions. But there were discussions behind the scenes where their views on biology, and genetics in particular, were expressed. Muller has given an interesting glimpse of this, showing how prevailing Marxist theories about science were taken seriously as guides in science. To give an impression of "the attitude of the leading party people," Muller described to Julian Huxley "a spirited argument" he had with Bauman and Iakovlev about "gene constancy." He was both amused and worried by their dogmatic dilettantism in scientific matters and by their facile explanation of the scientific beliefs of their opponents in terms of psychological and cultural factors:

> They think with Lyssenko that the gene is very labile/reflects its environment. Yakovlev accused me of deriving my scientific views of gene stability from a desire, for political reasons, to think the genes of all races alike and had some slogan to pigeonhole such a deviation in which scientific views are derived from social. On the other hand Bauman accused me of trying to carry over biological principles directly into the social level, and said we couldn't apply genetics to man that way. Yakovlev maintained that the genes of man had been changed by the environment of civilization and therefore primitive races existing today have inferior genes. But, he said, about three generations of socialism will so change the genes as to make all races equal. Just better the conditions and you better the genes. This, he said, represented the view in the highest official circles. But soon after that Bauman got into a hot dispute with Yakovlev on the question of race differences, denying what Yakovlev had said about this matter. Both men are highly self-opinionated and it was impossible for me to make an impression on their "ideology." Yakovlev kept harping back to the naive view: If the environment didn't change the gene, how could evolution have occurred?—as if this were an argument for Lamarckism. But they called their view Darwinian and I and Vavilov are accused of being anti-Darwinians because we believe in a high stability of the gene and in this change being "fortuitous."[15]

The December 1936 congress was a success for Lysenko. With respect to concrete agricultural issues, he met sharp criticism, which he was hard pressed to answer. But the result of the debate on basic theoretical principles of genetics was not as clear as the geneticists had hoped. His more substantial claims concerning genetics and breeding were severely criticized. But so was classical genetics in its pure-bred versions. Lysenko emerged as the leader of a rival and equally legitimate scientific school. And there was no doubt which of the "the two directions in genetics" was favored by the mass media, the party activists, and the political bosses. A number of established and well-respected scientists in genetics, breeding, embryology, and related fields gave Lysenko considerable support in his dispute with classical genetics. Meister, Sinskaia, Krenke, and M. M. Zavadovskii thought that Lysenko's Lamarckian ideas contained sensible hypotheses that deserved serious testing. Genetics was in a crisis and had no good reasons for a priori rejection. B. M. Zavadovskii even thought that Lysenko was much closer to the truth than geneticists such as Muller and Serebrovskii. Tsitsin supported Lysenko's intravarietal hybridization as a sensible reform in seed production.

There was no unanimity on the issues of the genetics debate within the top Soviet political leadership. The disagreement that Muller described between Bauman and Iakovlev over Lamarckism and human heredity is a sign of considerable confusion and potential deep disagreement. This situation may explain why Muller could entertain the hope that Stalin would respond positively to his proposal of a program for positive eugenics.

NOTES

1. On the history of Russian eugenics and its role in the genetics controversy see Adams (1990b).

2. For instance, Sapegin, at the discussion in the Leningrad Society of Marxist Biologists in January 1933; see chapter 5, "Theoretical Problems."

3. "Rech tovarishcha Stalina na pervom vsesoiuznom soveshchanii stakhanovtsev" (Speech of comrade Stalin at the first All-Union Congress of Stakhanov Workers), SS 4, no. 12 (December 1935): 8.

4. Agol was a young party member who had been active in criticizing Lamarckism in the late 1920s. Together with the human geneticist Solmon G. Levit, he had studied with Muller in Texas in 1931. Agol was a philosophical follower of Deborin and was accused of "menshevizing idealism" after Deborin's downfall in 1931. Agol had later organized and directed a genetics division at the Ukrainian Academy of Sciences. For his biography see Adams (1990c).

5. Carlson (1981) is the standard biography of Muller.

6. A copy found in Stalin's archive has been published by Iu. Vavilov, "Pis'mo Germana Miollera—I. V. Stalinu," *Voprosy Istorii Estestvoznanii i Techniki* 1 (1997): 68–76.

7. For a short chronicle of the congress, see *BV* 1 (1937): 20–26.

8. See, for instance, *Izvestiia*, December 24, 1936, p. 4.

9. Ibid.

10. The lectures of Serebrovskii and Muller are printed in the *Spornye Voprosy* as well as *SRSKh* 2 (1937). The remarks that Muller made, after Kol'tsov had read his lecture, were omitted, however.

11. On Kol'tsov's role in Russian biology and in the controversies over genetics see, for instance, Astaurov (1941), Adams (1980), Gaissinovich and Rossianov (1989), and Babkov (1989, 1992).

12. Letter from H. J. Muller to J. Huxley, March 9, 1937, Muller Archive, Lilly Library, University of Indiana, Bloomington. The letter was written on the train through Belgium and France after Muller had left the Soviet Union and was on his way to participate in the civil war in Spain.

13. Pavlov's small venture into genetics and how it was later mythologized is described by Krementsov (1997, 260–68).

14. "Ia obrashchaius' k Nikolaiiu Ivanovichu Vavilovu. Znaete li' vy gentiku, kak sleduet? Net, vy ne znaete, khotia, veroiatno, znaet eio luche, chem ogromnoe bolshinstvo zdec' prisutstvuiushchikh. . . . Vy malo zanimalis' drosophiloi i esli vam dat' obychnuiu studencheskuiu zadachu."

15. Letter from H. J. Muller to J. Huxley, March 9, 1937, Muller Archive.

Chapter 8.

Against Bourgeois Ideology in Science

With the great purges of 1936–1938 the despotism of Stalinism reached a climax. Among the groups most at risk were members of the Communist Party and military officers. In these groups the majority disappeared. The regime that developed under Stalin's leadership aimed for maximal loyalty to all decisions made at the top. The new Soviet constitution of 1936 was proclaimed as the most democratic in the world, and criticism was formally recognized as a core principle of democracy. But the system held all critical discourse—social, scientific, or political—within narrow bounds and strictly regulated debate. It must not be allowed to develop into opposition that could question the wisdom of the central leadership. Under this regime the economy and social structure of the Soviet Union was radically transformed, modernized in crucial respects. This revolution from above prepared the country for its key role in the victorious fight against Nazi Germany. An inhumane logic of calculated effectiveness was at work. Even the leading party members that were purged—sentenced to death or to long terms in prison—tended to accept the need for rigorous loyalty in a situation where the only important socialist country in the world was fighting a narrow battle for survival. This self-destructive mentality was analyzed by the Hungarian-born writer and disillusioned communist Arthur Koestler in his classic novel *Darkness at Noon*, published in 1940. The 1936–1938 purges triggered the first big wave of disillusionment with the great Soviet social experiment among Western liberal and left-wing intellectuals. The second wave of disillusionment came when the anti-Nazi alliance of the second world broke down and the cold war started in 1948.

The first Moscow show trial ended in August 1936 with death sentences for Grigorii Sinovjev, Lev Kamenev, former members of the Politburo, and fourteen other Bolshevik leaders. In January 1937 followed the second trial, where seventeen leading Bolsheviks, including Nikolai I. Muralov, the brother of the VASKhNIL president, received death sentences. In general, 1937 became a year of increasing terror and arrests. According to Russian historian Roy Medvedev, the second trial of January 1937 "marked the beginning of a new, much more terrible wave of repression" (Medvedev 1980, 162). On February 25, 1937, Bukharin was arrested at a meeting of the Central Committee, where he had come to defend himself against accusations of being a spy and saboteur. At the third Moscow trial in March 1938 he received his death sentence. In addition to these top leaders, hundreds of thousands of party members and other people were arrested and shot or sentenced to long terms of imprisonment.

A press campaign that linked classical genetics to political "traitors" was started by the agricultural newspaper *Sotsialisticheskoe Zemledelie* in the spring of 1937. Lysenko and Prezent in *Iarovizatsiia* accused their scientific opponents of being "Trotskyites-Bukharinites."[1] Bukharin's support for classical genetics against the Lysenkoists was now a liability. Scientists were fighting hard to uphold liberal ideals of scientific freedom and open criticism as well as traditional standards of argument and reasoning. In this situation the links to eugenics and race theories became fatal for genetics. It got caught up in the campaign against fascism.

KOL'TSOV IN THE AFTERMATH OF THE DECEMBER 1936 CONGRESS

Loyalty to "Darwinism" had become a touchstone for scientific truth in genetics. On January 7, 1937, less than two weeks after the

VASKhNIL genetics congress, Iakovlev applied it in a sharp attack on classical genetics. He spoke to a meeting of staff and authors at the state publishing house for agricultural literature.

Before Iakovlev spoke, the newly elected president of the Academy of Sciences, V. L. Komarov, gave an address in which he explained the great importance of the new editions of the works of Darwin and Timiriazev that were in publication. General knowledge of Darwin's work was scandalously poor. Komarov had asked one of his beginning courses at the university and found that only one person had read Darwin. But Komarov did not restrict his speech to such an uncontroversial critique of students. He went on to counter Vavilov's remark on Timiriazev's genetics at the December conference. Vavilov had warned that one should not take Timiriazev's criticism of Mendelism as literally as Lysenko did: Timiriazev was a great polemicist and in the heat of a debate he had the ability to make cutting but not always "objective" remarks (Vavilov 1937, 49). But it was a mistake to look upon Timiriazev as old-fashioned, said Komarov. He was not only an able polemicist but also an outstanding plant physiologist with an unusually broad comprehension of biology, a man who understood "the essence of things" (Komarov 1937, 27–28).

Iakovlev's speech was an extended polemic against "anti-Darwinism" in contemporary biology. He found classical Mendelian genetics to be incompatible with true Darwinism. None of the geneticists dared openly to reject Darwin, admitted Iakovlev, but their attitude was clear from the theories they proposed. In particular, he pointed to the "immutability of the gene" as incompatible with biological evolution. Vavilov's theory of homologous series, for instance, built on this misconception, as did the distinction between genotype and phenotype, which the geneticists were so fond of.

Iakovlev also linked classical genetics to eugenics. When a theory that was so obviously false had such great success, it had to be for political rather than scientific reasons, he argued. Mendelian genetics denied progressive biological evolution and was therefore a suitable basis for the idea of superior races. He quoted Timiriazev's alleged characterization of Mendelian genetics as "a clericalist and nationalist" intrusion into science. The fascist use of neo-Mendelism for political purposes "confirms this evaluation of Mendelism by Timiriazev," declared Iakovlev. The issue was whether one should further the development of genetics from the point of view of evolution or let it be converted into a "maidservant for the department of Goebbels" (Iakovlev 1937, 24–25). Apparently some people in the party apparatus at this point had developed strong views about what a true genetic science had to be like.

The old liberal Kol'tsov was not scared to silence.[2] Soon after the December conference he wrote to the president of VASKhNIL, Muralov, criticizing the way the conference had been organized and conducted. It had turned into a destructive attack on the science of genetics. Kol'tsov was worried about the effect that misleading press reports would have on the teaching of genetics throughout the country. He asked Muralov to support the publication in *Pravda* and *Izvestia* of "extensive articles, written by genuine geneticists in defense of their science." It was also necessary, in the nearest future, to organize a meeting of VASKhNIL academicians with Iakovlev and Bauman of the Central Committeee, wrote Kol'tsov. "Perhaps it will be possible to take measures that can save science," his letter ended.[3]

Muralov organized a special meeting of the presidium of VASKhNIL to discuss Kol'tsov's letter. In addition to the members of the presidium, Kol'tsov and Konstantinov were present. The majority of the presidium, consisting of Muralov, Meister, and Margolin, resolved that Kol'tsov's letter gave an "incorrect evaluation" of the results of the discussion at the congress. In their view the discussions had stimulated "the interest for questions of genetics and selection in a broad popular audience" and "brought forth a number of claims that can be tested experimentally." The minority, consisting of Vavilov and M. Zavadovskii, abstained with the motivation that "at the present stage the negative results are most evident," but "in the future the staging of experiments and analysis of controversial questions will prove useful for the interests of science."[4] But there was no real discussion at this meeting. The issue had apparently been decided in advance. First, Zavadovskii and Vavilov spoke about Kol'tsov's letter. Then the others briefly stated their opinions, and a vote was held.⁵ Two months later Margolin, at the March *aktiv* of VASKhNIL, with great indignation, accused Vavilov of having said that the presidium was "playing with fire."⁶

In a long letter of reply to Kol'tsov, Muralov attacked Kol'tsov's elitism and his former eugenics. The problems of genetics had to be discussed in the "wide circle of scientists and production workers" not only in a "closed circle of narrow specialists." It was well known from the history of science, argued Muralov, that the large advances were made by nonspecialists. A new epoch in medicine was created by the biochemist and crystallographer Louis Pasteur, and Mendel himself was a specialist in theology. Kol'tsov wanted to let twenty-seven academicians decide and exclude all the others who were building Soviet science. Muralov found it impossible to agree to such an undemocratic procedure.

THE VASKhNIL *Aktiv*, MARCH 26–29, 1937

A special expanded meeting of the VASKhNIL presidium, a so-called *aktiv*, was called toward the end of March. A new constitution had just been introduced in the Soviet Union, claimed to be the most democratic in the world.⁷ A special plenary session of the Central Committee, from February 23 to March 5, had discussed the implications of this new constitution. Among other things, Stalin had declared a sharpened class struggle at the ongoing stage of the building of socialism. The meeting at VASKhNIL was one in a wave of parallel meetings throughout the country called to discuss the resolutions from the Central Committee session. The *aktiv* was to pursue the struggle on the "ideological front of our agricultural science," as Muralov expressed it in his introductory speech.⁸ Kol'tsov's letter was a welcome handle to reveal the lingering bourgeois ideology among the academicians. His eugenic writings from the early 1920s were effectively used as a whip to force geneticists into submission.

The *aktiv* was mainly a local meeting of VASKhNIL members and employees living in the Moscow region. The meetings were attended by perhaps forty to sixty people and discussions were much more personal and direct than the large public meetings of the December session. The verbatim reports from this *aktiv* give a much clearer picture of the internal VASKhNIL struggle between politicobureaucratic and scientific views and attitudes. On the one side were Muralov, Margolin, and Bondarenko, on the other Kol'tsov, Prianishnikov, and Vavilov. Though all six were formally academicians (academy members) and thus fully qualified to judge in scientific matters, they fell clearly into two camps: scientists and politicians-administrators. The tug of war took place in an atmosphere of populist revolt against academicians and other scientific authorities, an aspect of Stalin's revolution from above. But the political leadership also felt insecure, threatened by the party purges. Prezent and Lysenko were fishing in troubled waters. They lashed out with aggressive and rude criticism toward both sides. The meeting was characterized by ideological frenzy and a chaotic situation with respect to administrative and scientific questions.

The frenzied ideological atmosphere that had now developed in the genetics debate was a hard strain on many scientific members of VASKhNIL. From the politicoadministrative side, it was complained that so few of the academicians were present and active in the debates. It had more the character of a "passive" than an *aktiv*, said the witty and ironical polemicist Prezent.[9] Meister did not participate in the discussions—perhaps not so surprising since his base was in Saratov. Neither did Serebrovskii, though he was partly present, listening passively. Kol'tsov in his speech complained about the tone that had come over the genetics debate. Some printed articles were of such a character that the responsible editors ought to be penalized. Dimitrii Prianishnikov had written a letter to Muralov about this. Muralov replied from his seat: "A reproof was pronounced." Kol'tsov himself had been ordered by his doctor to take a rest. At the resort he had met five other academicians suffering from the strains of the December session. Margolin interrupted: "Lysenko was also suffering in a sanatorium."[10]

Muralov in his introductory speech proclaimed that the split which Kol'tsov's letter had produced in the presidium of VASKhNIL showed that "we have not organized our cadres to fight the bourgeois worldview." But now, thanks to this letter, we "have come to know our academicians, have come to know their worldview." To demonstrate these harmful worldviews, he gave two quotations from articles by Kol'tsov in the *Russian Journal of Eugenics* in the 1920s. The first one appears rather uncontroversial and innocuous. It said that epidemic diseases such as cholera, smallpox, or tuberculosis could be seen as a means to select away weak constitutions, and it was thus "beneficial for the physical health of the race." The second was more provocative, pointing to the possibility of human breeding:

> If the laws of Mendel had been discovered a century earlier, at the time of Russian [landed] proprietors and American slave-owners, who had power over the marriages of their serfs and slaves, they could have obtained special desired types of humans, by applying the theory of heredity.

Kol'tsov objected from his seat that the quotations were taken out of their context and misinterpreted. But Muralov maintained that such views expressed bourgeois or even fascist ideology and were unacceptable in the new Stalin constitution.[11]

On the second day of the *aktiv*, Kol'tsov valiantly defended himself in a long speech. He complained that the text of his letter had not been distributed so that everybody could read and judge its content. It was a misunderstanding that he was against broad public discussions. What he had criticized was the way the presidium had handled the VASKhNIL December session and its public relations, said Kol'tsov.[12] For instance, it had failed to do anything to correct the very misleading and tendentious press reports. But Muralov continued to be quite unwilling to promote any popular newspaper articles on genetics. Meister's concluding summary had been published in an agricultural journal, and that was enough.[13]

With respect to his eugenics, Kol'tsov claimed that Muralov had been deceived by quotations out of context. It was easy to pick, for instance, from *Pravda* or *Izvestia* phrases that in isolation could appear counterrevolutionary. In its proper context, his own example of hypothetical human breeding was used to support the view that such breeding was technically feasible under special circumstances, but that it was unacceptable.[14]

Kol'tsov's valiant defense did not save him from becoming the whipping boy of the *aktiv*. B. Zavadovskii said ominously that Kol'tsov's criticism of the VASKhNIL presidium had "a definite political character, a definite political color."[15] Margolin presented new quotations to prove that Kol'tsov had in fact been a eugenicist. For instance, Kol'tsov had written:

> Due to some extra difficulties and complications eugenics differs from animal breeding, but we can imagine conditions when human nature could be improved with the same methods that contemporary breeding uses to improve stocks of animals.

Margolin also quoted Kol'tsov on the eugenic effects of war. He had said that civil war might be worse than international war because it tended to eliminate a larger proportion of the elite of a country.

The critics were infuriated by Kol'tsov's unwillingness to perform self-criticism and repent what he had written. Instead, he claimed that at the time his writings had been quite legitimate and innocuous. He found no reason to take back a word of what he had written. Margolin found it highly abnormal (*nenormalnyi*) that a man with such dubious attitudes and opinions should present himself proudly as a modern Galileo defending science.[16]

Margolin adressed his polemics directly to Vavilov: "Nikolai Ivanovich, you saw how Serebrovskii was forced to retreat" (*kak Serebrovskii razbili*). "Why did you not tell us in the presidium (at the December session) that Kol'tsov was of the same kind? You must have known." "I never read this nonsense," Vavilov replied somewhat evasively. "We did not read, you did not read, but you knew him, and nevertheless you recommended him as a member of the Academy of Sciences," continued Margolin. Said Vavilov: "He is a great scientist." Replied Margolin: "You heard what this great scientists writes."[17]

Eugenics and racism at the planned international genetics congress also worried Margolin. He pointed out that Kol'tsov and Vavilov were still members of the German Society for Genetics, a society that was highly influenced by Nazi eugenics and racism.[18] During his speech, later on the same day, Prezent picked up this theme. He first confirmed with Vavilov that he would vote for Kol'tsov as a member of the Academy of Sciences, and then continued to ask if he would "let Bauer, great scientist and Hitler facist, into the academy? I believe the question is such that the answer should be completely clear." Vavilov said: "Bauer died three years ago." Replied Prezent: "The matter is that Vavilov has not learned to give priority to political judgment" (*politicheskuiu rastsenku postavit' vo glavu ugla*). Hitler had purged the German genetics society of unwanted elements, Prezent told the audience, when were Vavilov and Kol'tsov going to leave it?[19]

In his own talk, Vavilov answered that the German genetics society was an important international scientific society with members from many countries. In view of the coming international genetic congress in Moscow, it was important to keep up these contacts. By terminating his membership in protest, he could cause a break of diplomatic connections that would harm the congress. The foreign minister Litvinov had personally proposed that he should not hurry in sending his resignation. However, he had no personal interest in this and was willing to resign at any moment, stressed Vavilov. "I acted with caution in this case exclusively in the interest of Soviet science and the Soviet Union."[20]

The violent criticisms leveled at Kol'tsov gave Prezent good reason to triumph. The ideological position of Kol'tsov should be well known. Neither Margolin nor Muralov are novices in "questions of politics and Marxism," said Present. They surely read Marxist literature in 1931 when his paper about the ideological deviations of geneticists Kol'tsov and Filipchenko was published in *Under the Banner of Marxism*. Prezent complained that Kol'tsov, who had spoken the day before, was not now present to hear the criticism of his views. He nevertheless continued the reading of quotations from Kol'tsov's eugenics articles and books of the 1920s.[21] Under the circumstances, this gave him good response from the audience and nobody spoke up to defend Kol'tsov's stand on self-criticism.

Besides this ideological bashing of geneticists, there was also a more balanced and sober debate on questions of administration and steering of research. Muralov worried about the lack of means to see to it that the institutes actually followed the instructions of the presidium. He used as his example the search for varieties of wheat resistant to rust. Plant diseases and resistance was one of Vavilov's specialties. Muralov described what he saw as a lack of purposeful, determined, and well-planned attempts from the scientists to solve problems of agriculture.[22]

Vavilov in his own speech did not respond to Muralov's example, wheat rust, but repeated in sharp formulations the criticism of bureaucratization and the lack of scientific judgment from the reorganization in 1934–1935. According to the Soviet government resolution of 1935, VASKhNIL had two main defects: first, the lack of an effective "collective of academicians," which could unite them in their scientific work and link it to practical agriculture. Second, a bureaucratic "system of steering" (*upravlencheskaia sistema*) that was no good for scientific research. All the selfless efforts of Muralov, as well as the entry of a new academic secretary with strong practical administrative qualifications, had only strengthened this harmful "spirit of steering" (*upravlencheskii dukh*). Vavilov put the Academy of Sciences, where he was also a member, up as a counterexample, a real scientific academy where the scientists played a central and productive role in governing the activities of the organization.[23]

Prianishnikov joined Vavilov and Kol'tsov in criticizing the lack of administrative efficiency and scientifically competent leadership. He argued that VASKhNIL had failed in carrying out its main task of planning and steering scientific research, the "strategic role," as it is often called in present politics of science. This task was suffering due to the neglect of theoretical scientific knowledge, and short-term practical objectives were favored. According to Prianishnikov's diagnosis, the central bureaucracy of VASKhNIL lacked contact with the research activity—the bureaucrats were not interested in or did not understand the content of the research. Sometimes the presidium gave instructions and then simply forgot about them, said Prianishnikov. This had happened, for instance, in the case of hydrolyzing straw from grain growing. An interruption from the audience reminded Prianishnikov that a special brigade had been formed at the VASKhNIL secretariat to work on this topic. Prianishnikov replied: "But the brigade did not come to us." New interruption: "But you yourself spoke in the VASKhNIL." "Yes," said Prianishnikov, "we came to explain our research to a meeting at which Bondarenko presided." But he "was occupied with different matters at this time, and did not hear one word of the talk." Muralov's introductory speech for the meeting of the *aktiv* had contained much important criticism, concluded Prianishnikov, but "the positive part was absent.... Besides criticising it is necessary to point out the basic tasks in the restructuring of the academy."[24] Prianishnikov's criticism of the regime in VASKhNIL was low-key but precise and pointed. The implication of his remarks was that the principle of "unity of theory and practice," as it had been implemented in the reorganization in 1934–1935, had had the opposite of the intended effect.

In his closing speech, Muralov challenged Vavilov on the problem of steering scientific research: he had seriously attempted to promote the production of rust-resistant wheat varieties through contacts and instructions to the research stations, and Vavilov seemed simply to call this bureaucratic interventionism (*upravlenchestvo*). "Should I just keep away from it?" Muralov asked rhetorically. "Absolutely," replied Vavilov from his seat. "Is it not the task of the presidium to organize research so that new varieties can be gotten as quickly as possible," retorted an agitated Muralov. "You call this bureaucratic interventionism, but we call it organization of research." "You should take advice from the best specialists," answered Vavilov in a more accommodating mood, mentioning a couple of names. "But you did not say that in order to produce plants that are resistant we should invite Lysenko and Tsitsin," continued Muralov. Vavilov did not reply. "I will fight to get new varieties as quickly as possible," Muralov concluded the exchange.[25]

A notable feature of the *aktiv* was the restraint on criticism of Lysenko. When mentioned, he was handled very circumspectly by both main sides in the debate. Lysenko's stature as an icon of the new revolutionary type of science, "proletarian science," together with the personal favor he enjoyed with Stalin, explains much of their restraint.

Prezent was dissatisfied with the negative attitude of the academy toward the vernalization of seed grain. Bondarenko had been very negative. Even Muralov had not been quite neutral at the December congress, said Prezent. How could it, for instance, come about that a troika consisting of Kostov, Konstantinov, and Listsyn had been named to plan experiments to sort out the controversial questions? And Margolin, during a visit to Odessa, had not even bothered to look at the experimental fields. [26]

Lysenko spoke about the vernalization of seed grain and intravarietal crossing in his usual unsystematic and anecdotal way. He repeated his complaints about Konstantinov's criticism of vernalization. He said that while the academy had been very little interested in hearing about his results, the agricultural daily and the agricultural publishing house had been much more responsive. According to Lysenko himself, the idea of using *kolkhoz* workers to implement a large-scale scheme of intravarietal crossing in order to renew the seed was held up by Meister in particular. Meister demanded experimental tests before Lysenko's new scheme was applied on a grand scale, while Lysenko wanted immediate action.

"Look why I have come to this meeting," said Lysenko to Muralov, "I am not giving up my plan to carry out intravarietal crossing on between fifty and seventy thousand collective farms."[27] This year "I need 300–500 thousand scissors," explained Lysenko. But to make these scissors, five tons of steel were needed, which he did not have. "Margolin might say that this is not a matter for the academy," commented Lysenko, but I know "that without such things there is no science" (*shto nauki bes takikh del net*). He would go over the head of the academy if necessary: "If you do not help me, I will go to the Ministry of Agriculture, breaking down their doors, if I do not get it in the ministry, I go further and I will get it. . . ." This threat concluded Lysenko's speech, which the audience rewarded with applause.[28] The thinly veiled suggestion was that he had good contact with Stalin, who would help him out if nobody else did.

Vavilov's response to Lysenko's scheme of intravarietal crossing was highly accommodating, though in vague and veiled words. "This is a very big question which demands decisive action," said Vavilov. And the academy had always been favorable toward Lysenko and tried to meet his wishes. Vavilov suggested that a commission headed by the president of the academy should visit Lysenko to discuss the matter. Then an appeal could be sent to the ministry and other government agencies about "appropriate means" (*sootvetstvuiushchikh meropriiatiakh*).[29]

Muralov was similarly keen not to antagonize Lysenko. He said in his concluding speech that an agreement had been reached with Lysenko that a commission with himself as chairman and Vavilov, Lysenko, and Meister as members would review the question of intravarietal crossing. If he had still been vice minister of agriculture, said Muralov, he would immediately have given Lysenko his five tons of steel, but the president of VASKhNIL did not have such means at his disposal. Muralov also tried to appease Lysenko on the question of vernalization: "I was never against vernalization. . . . I welcome and support every new means (*priiom*) which gives an effect, not over ten years but even for just one or two years."[30]

Prezent grasped the situation well when he maintained in his speech that "the fundamental question for the academy is the question of scientific politics. . . ."[31] Margolin, in spite of his sharp criticisms of Prezent's polemical behavior, also emphasized this basic principle. He started and ended his speech by admitting that the criticisms for administrative confusion and delays in the presidium were right and deserved. With the lack of administrative resources, it could hardly be otherwise. But much graver was the criticism for "lacking recognition of the importance of the ideological struggle and inability to pursue it in the right way."[32] In other words, the primary task of the politics of science was to ensure that science served the political ends of the people and its government.

CANCELLATION OF THE INTERNATIONAL CONGRESS OF GENETICS[33]

In the spring of 1932 Vavilov had made efforts to get the Seventh International Congress of Genetics to the Soviet Union. A strong Soviet showing at the sixth congress to be held in the United States in August 1932 would be a good argument. In March Vavilov wrote to the then minister of agriculture, Iakovlev, urging that a substantial and representative delegation be sent, including senior scholars as well as young apprentices. This would show the achievements of Soviet research and demonstrate the world that the Soviet Union was well prepared to host the next International Congress of Genetics. Many scientific institutions as well as individual scholars had received invitations to the congress in Ithaca, explained Vavilov. It was the time to decide who was to participate so they could properly prepare their lectures, exhibits, and so on.[34] However, the political leadership was not eager to support foreign travels for Soviet scholars, and Vavilov had to represent his country almost alone. A number of Soviet geneticists had prominent places on the program but did not turn up.[35] And Sweden was chosen for the next congress.

But in the spring of 1935 the Swedes declined the invitation, and Vavilov seized the opportunity to propose the Soviet Union as host for the upcoming congress. After preliminary contacts in the spring and early summer,[36] Vavilov in August 1935 could write to the chairman of the Permanent International Committee on Genetics, the Norwegian human geneticist O. L. Mohr, that "we have obtained permission from the Soviet government to invite the Seventh International Congress of Genetics to the USSR" "Our group of geneticists is young," admitted Vavilov, "and, of course, we understand that we shall have to pass an examination." He counted on the American geneticist H. J. Muller, then working at the Institute for Genetics of the Academy of Sciences in Moscow, to "help us a great deal."[37] Mohr was a good friend of Muller's since 1918–1919 when he spent a year in the group of T. H. Morgan at Columbia University. At first, Muller had advised against the idea of an international congress in the Soviet Union. He thought the Soviet geneticists did not have sufficient experience. To undertake such a big task in 1937 would be premature. It would also take much energy away from the research work.[38]

The official invitation from the Soviet Academy of Sciences was promptly circulated by Mohr to the members of the international committee with his recommendation that it should be accepted.[39] By December 1935 the international committee had agreed[40] and a preliminary program and a Soviet organizing committee had been proposed by the institute of genetics and approved by the presidium of the Academy of Sciences.[41] The organizing committee was approved by the Soviet government in February 1936 with Muralov as president, Vavilov and Komarov as vice presidents, and Solomon G. Levit as general secretary. The nine other members were the academicians Gorbunov, Lysenko, Keller, Kol'tsov, Meister, and Serebrovskii, and professors Karpechenko, M. S. Navashin, and Muller.[42]

A conflict over human genetics and its medical applications was now emerging as a central issue in the internal Soviet debate over the international congress. H. J. Muller was chairman of the program committee and eager to give human genetics, including race questions, ample room on the program. He had a double purpose in mind. On the one hand, he wanted to criticize Nazi eugenics and reveal its lacking scientific foundations. On the other hand, he wanted to help build a solid scientific basis for a genuine socialist eugenics. Muller had also worked to make Levit general secretary of the congress. Solomon G. Levit was a human geneticist and a party member since 1919–1920. He had visited Muller with a Rockefeller stipend in Texas during 1931, together with I. Agol. From 1932 Levit was director of the Medico-Biological Institute that in 1935 was renamed the Maxim Gorky Scientific Research Institute of Medical Genetics.[43]

A common meeting of the program and organizing committees produced a draft program in March 1936. Out of seven plenary sessions, one was to be called "Questions of Plant Genetics and Breeding," with Lysenko and the Swedish plant breeder and geneticist Herman Nilsson-Ehle among the speakers. Another was to be called "Human Genetics and Its Application to Medicine, Anthropology, and Psychology." Three speakers were proposed for this seventh and last session: Lancelot Hogben and J. B. S. Haldane from England, and Levit from the Soviet Union. Three alternate speakers were also listed: Otmar von Verschuer from Germany, Kristine Bonnevie from Norway, and Lionel Penrose from Great Britain.[44] A proposal from the program committe two weeks later had a few changes. Julius Bauer of Austria was added as a speaker, and Eugen Fischer of Germany had replaced Verschuer as an alternate.[45] A late April draft of the plenary sessions listed Huxley, Mohr, and Levit as speakers with Hogben, Bonnevie, and the American geneticist H. S. Jennings as alternates. There were no Germans anymore.[46]

In a September 1936 memo from the organizing committe the antifacist purpose of the session on human genetics was stressed. The Russian speakers were to concentrate on two tasks: "a revealing theoretical criticism of the pseudoscientific character of fascist theories on race and eugenics" and a presentation of Russian work in medical genetics. For the foreign speakers it was proposed that they should primarily speak "against fascist race theory and fascist eugenics." As possible candidates for this task were mentioned Julian Huxley and Lancelot Hogben (England), Mohr (Norway), and Jennings (United States).[47] All these were known for their public criticism of radical eugenics policies, especially the kind that were practiced in Germany after 1933.

But there was disagreement on the exclusion of the Germans. Muller wrote to Mohr in October 1936 that one of six plenary sessions of the genetics congress was planned to be called "Race Theories and Related Subjects." The program committee was considering Huxley for first speaker, "a German such as, for instance, Fischer"[48] for the third, and Levit for the fourth and last. For the second speaker they wanted Mohr himself.[49] It is noteworthy that Muller and the program committee were still considering the physical anthropologist and eugenicist Eugen Fisher. Their desire for an open and politically impartial debate of scientific issues was still resisting the pressure for political correctness.

Political uneasiness about the planned international congress in the context of the intensifying internal Soviet discussions over breeding and genetics was growing. A special commission reviewed the preparations and gave a rather negative report in early October: among the foreigners who had signed up for the conference, there were "many fascists." The chairman had attached a list of people he knew to be fascists. It was also noted that the "question about speakers in the section of human genetics/race theory has not yet been decided," and that there was only one communist in the organization committee, namely, Levit. Furthermore, the practical preparations for the congress were found to be lagging far behind schedule. Without a radical change in its working tempo, preparations would not be finished in time.[50]

A note to Gorbunov, still permanent secretary of the Academy of Sciences, from a worker at the genetics institute of the academy, argued that Soviet genetics was not yet well prepared for such a broad international presentation as the congress would be: the ongoing "controversies between Lysenko and the followers of so-called formal geneticists" had demonstrated that so far "we do not have sufficiently solid work in theoretical genetics." As a consequence, there would be an "excessive" proportion of foreign speakers on basic theoretical questions. In many cases "we will have to give a Soviet rostrum to fascist speakers." Furthermore, the new building of the institute and its greenhouses would hardly be ready in time. These circumstances "might produce an unfavorable impression" that science develops more poorly in a socialist than in a capitalist society.[51]

The pressue against human genetics increased through 1936. The Central Committee condemned intelligence testing. Arnost Kol'man, one of the leading party ideologues of science, on several occasions denounced Levit's Institute of Medical Genetics for fostering fascism. This campaign led to a public meeting that condemned levit's human genetics as fascist doctrine on November 13, 1936, one and a half months before the VASKhNIL congress on genetics and breeding (Adams 1990d).

On November 17, the Soviet Council of Ministers decided to cancel the congress. The Academy of Sciences quickly sent a letter to the prime minister, Molotov, asking for a postponement to 1938, rather than a permanent cancellation. Such a postponement of one year would be the best solution, "fully explainable by the preparations for the twentieth anniversary of the October Revolution and the international geological congress to be held in 1937." A complete cancellation "might produce unwanted interpretations and statements not in the interest of the prestige of Soviet science."[52]

The cancellation was not immediately communicated to the international committee and its president, Mohr. On December 8, 1936, Muller sent a short letter to Mohr saying that there would be no congress in Moscow the following summer. Muller was writing quite unofficially, not as "a member of the program or organizing committees, since it is my understanding that the latter no longer exist."[53] Similar disturbing information reached Mohr through other channels. In mid-December he received a telegram from the American geneticist Ralph Emerson that the congress in Moscow had been canceled. The information was based on a notice in the *New York Times* on December 14. This notice said that a serious schism had developed among leading Soviet geneticists, that some of them—including members of the organizing committee of the international congress—had been arrested, and that the congress had been canceled. Mohr immediately telegraphed the organizing committee in Moscow. On December 26 Muralov replied that the congress was not canceled but only postponed.[54] The *New York Times* had exaggerated. Vavilov telegraphed the newspaper to complain about its "slanderous reports" of his arrest.[55] But it was also clear to Mohr that Soviet genetics was having severe political problems, and an international congress organized in Moscow could be in deep trouble.

The international congress had become an important issue in the genetics controversy. This was clearly demonstrated in the VASKhNIL discussions in late 1936 and early 1937. As Muller's biographer writes, "The prestige and publicity of an international congress would permit bona fide genetics, theoretical and applied, to reach the Soviet citizenry and it would show how insignificant the Lysenko movement was in the eyes of the world" (Carlson 1981, 227).

Muller's efforts to make the Permanent International Committee on Genetics uphold the decision to hold the seventh international congress in Moscow and accept the postponement to 1938 did not succeed, however. In the end, the majority of the committee, led by its chairman, Mohr, decided that the congress should be rescheduled for Edinburgh in 1939.

But how open was the Soviet situation in 1937? Could a more accommodating attitude from the international scientific community have helped Soviet geneticists withstand the pressure from Lysenkoism? Was the political situation of Soviet genetics so bad already by early 1937 that there was no outlook for a scientifically successful congress in Moscow in 1938 or 1939? How well was the international committee under its chairman, Mohr, able to judge the situation with the spotty and unreliable information it received?

Mohr was a left-wing liberal with a strong belief in the mutual link between political and scientific freedom. He was not involved in party politics but voted Labor. Mohr was trained as a medical doctor and had become a specialist on human genetics. Among geneticists he was also one of the staunchest opponents of eugenics policies (Roll-Hansen 1980). By the middle of the 1930s, Mohr was a strong antagonist of the Nazi regime in Germany. In a letter to his friend the American geneticist Leslie Dunn, Mohr tells how the rector of the University in Oslo had planned to participate in the 550th anniversary of Heidelberg University, but was prevented by Mohr's intervention.[56] By the end of 1936 he was also highly critical of the increasingly repressive and brutal dictatorship in Russia. Conversations with H. J. Muller in Copenhagen earlier in that year, as well as the executions of Sinovjev and Kamenev, had confirmed his fears. For Mohr the antiliberal and antiscience features of dictatorships were being revealed both in Germany and in Russia:

> Without freedom of utterance real humanity may not exist. The main achievement of science is that it has removed the fear of the unknown forces of nature. But when freedom is suppressed this fear is replaced by fear among individuals and among nations.[57]

On January 7, 1937, Mohr sent a letter to Muralov, as president of the local organizing committee, with copies to Vavilov and Levit. He referred to the article in the *New York Times* and rumors that Vavilov and others had been arrested. "Until I am otherwise informed by you I will not believe this information," wrote Mohr diplomatically. "But it is under all circumstances urgently needed that you send me immediately detailed information on the situation." Mohr needed to answer the questions coming from the members of the international committee and to make his own judgment of the situation. He needed to know whether cancellation was the more or less inevitable outcome of current Soviet developments:

> . . . if it should turn out that there is a real, though less alarming basis for the above mentioned statements [in the *New York Times*, and so on] I think we may as well immediately face the situation that cancellation, and not only a postponement will be the final result.[58]

In late January 1937 Mohr described the situation in a long letter to Dunn. The second Moscow trial had started, and Mohr sighed: "Isn't it fantastic and incredible! For me who had really believed in a sound development in [the] USSR." A letter from Vavilov written January 4 and received two weeks later had given a relatively neutral description of the December genetics congress in Moscow. Vavilov had also explained that a difficulty was that the Soviet government did not want to include the "racial problem" in the program of the congress. A letter signed by a number of prominent American geneticists and demanding that a discussion of the racial problem should be included in the congress had been received by the organizing committee about seven months earlier. This letter had been published in Soviet newspapers, and there had appeared "undesirable comments in the German press," wrote Vavilov. But Mohr was still impatiently waiting for more precise information about the reasons for the postponement and the events around that decision, he explained to Dunn.[59]

Toward the end of February, Mohr received a letter from Muralov and Vavilov that gave the following clearly incomplete explanation of the postponement:

> A number of institutes and scientists requested this as they wished to be better prepared for the congress and it was therefore decided

to change the date of the congress to 1938. We have now obtained permission from the government to hold the congress in Moscow at that time.[60]

More important, a number of the bulletins from the second half of the December genetics congress were enclosed. These were translated, and Mohr could read for himself how the Lysenkoists argued. In his reply to Muralov, Mohr noted the politically sensitive nature of genetic science and mentioned that "quite a few members of our committee have already asked whether it would not, under the present conditions, perhaps be better to have the next congress in a country that is not in the foreground of political attention."[61]

In late March a special commission headed by academy president Komarov had taken over the responsibilities of the organizing committee in Moscow. In a telegram to Mohr in late March, the invitation for August 1938 was repeated, and he was officially invited to visit Moscow to discuss "the best possible way to prepare the congress." The commission had also decided that "race theory was to be excluded from the program of the congress."[62] By this time Mohr doubted the usefulness of a visit to Moscow and did not immediately respond to the invitation. He was playing for time. As he explained some months later to his friend Dunn, "Time was necessary in order to get a real idea of what was happening in Russia."[63]

On April 8, 1937, Mohr sent a "memorandum" to all members of the international committee, including the Soviets. He described briefly the events since December and enclosed transcripts of his official correspondence with the organizing committee in Moscow. His opinion, based on the bulletins from the December genetics congress, was that "rather deep-seated differences on very fundamental genetic principles have arisen among leading Soviet geneticists, differences which in the writer's opinion are of such a nature that they may be expected to make themselves felt during an international congress." With careful diplomacy, in order not to embarrass his Soviet colleagues, Mohr expressed his most serious concern for a free scientific debate. He also informed the committee that he was investigating alternative places for the congress. The memorandum was mainly for information, but he also asked the members for "their general, noncommittal opinions in the matter. I need this in order that my proposed personal visit to Moscow shall be of full use."[64]

In early April, Vavilov wrote optimistically to Muller:

> The government and the Central Committee are now definitively in favor of the congress. Molotov himself spoke by telephone with Komarov about the congress, about the program, about the organizing committee. For an hour they spoke frankly about all questions. It was quite an unusual event as you will understand. Molotov and Litvinoff[65] are now engaged in arranging the congress to be held in the USSR. . . . So now we have started a campaign for the Congress, and we ask you to help us in this matter. Of course, you understand that it is very important for the benefit of genetics to have it in the USSR.[66]

On the same day Vavilov also wrote to the Finnish geneticist Harry Federley about the congress. Once more, he wrote that the race question was the main reason for the postponement. In January he had written to Mohr that personally he had been "somewhat in doubt" of the need to include this problem in the program. Now he said that he had been against it. And he added, writing in German: "As the question of the race problem does not exist in our country, our government has decided against its being included in the program of the congress."[67] The last statement indicated that the Soviet government was prohibiting any lectures and public discussion of questions concerning human race differences or even human genetics at the congress. This was hardly acceptable to the international committee, as Vavilov must have been well aware. By this time he had probably lost all real hope in the congress, but was still keeping up appearances for the sake of internal Soviet politics of science.

In early May 1937 the Academy of Sciences continued its efforts to revive the congress project by naming a new organizing committee headed by Komarov, including Vavilov, Muralov, and Lysenko as vice presidents and Meister as general secretary. Levit and Kol'tsov were out. A set of five plenary sessions was sketched with no human genetics. Komarov was named as Soviet speaker in a plenary session called "The Evolutionary Theory of Darwin and Contemporary Genetics." Other proposed Soviet speakers for the plenary sessions were Vavilov, Lysenko, Meister, Tsitsin, Muller, and Levitskii. The Soviet speakers were obliged "no later than November 1937, to present the theses of the lectures to be read by the organizing committee."[68] The protocol of these decisions was officially sent to Bauman, head of the science section of the Central Committee of the Communist Party.[69] It was clear that political control was tightening.

Muller was still loyally working for a Moscow congress. In mid-May he wrote to Mohr arguing strongly that abandoning the Moscow congress would make the situation worse for the Soviet geneticists. They badly needed the congress in their fight for survival. Under the pressure of political developments and the attacks against genetics, Muller had left Moscow in March to participate in the International Brigade in Spain. After two months in Spain he was on his way to the United States, where he would spend another two months before returning to Moscow "for a time at least."[70]

At about the same time, Mohr informed the Soviet organizing committee that he had postponed his intended visit to Moscow. He explained diplomatically that he had deemed it advisable to come somewhat later, hoping that time might aid in straightening out to some extent the rather profound differences of opinion that judging from bulletin on the session on genetics and selection held in December, had developed among leading Soviet geneticists.[71]

In the same letter he mentioned new rumors about arrests, this time of Kol'tsov and Serebrovskii. An immediate reply from Vavilov told that this was "entirely wrong": "Both are continuing their work without interruption."[72]

A resolution from a meeting of the organizing committee in mid-June complained that "the libelous attacks on Soviet science" in the international press were "completely false and of provocative nature."[73] A letter to Mohr pressed for a decision about the congress and again urged him to visit Moscow.[74] He also received a letter signed by a number of the most prominent Soviet geneticists pleading for the congress to be held in Moscow. This letter once more informed Mohr that Serebrovskii and Koltsov had not been arrested, despite the attack on them that Prezent had published in *Sotsialisticheskoe Zemledelie* on April 12. It also maintained that all Soviet scientists were free to express their scientific opinions and that conditions were favorable for a congress in Moscow in 1938.[75]

By mid-July Mohr had cleared his mind of the last doubt. On July 21 he sent a new memorandum to the international committee informing the members that the reaction to his earlier memorandum of April 8 had been clearly in favor of rejecting the new invitation to Moscow in 1938 and moving the congress to another country. Mohr had received twelve answers, out of which eight were definitively negative toward Moscow. Furthermore, he had established that the British Genetics Society was willing to take responsibility for a congress in Edinburgh in 1939. Mohr pointed to the special political sensitivity of genetics and the tense political situation in Europe, mentioning that several members of the committee had advised that the congress be held in a country where "this situation is least likely to be felt." He then asked the members for a vote on two alternatives, Moscow 1938 or Edinburgh 1939.[76] In September Mohr informed the Soviets and the international committee about the result. Out of fifteen members, three had voted for Moscow and nine for Edinburgh.[77]

Writing to his friend Dunn in August 1937, Mohr defended and explained the course of action that he had taken after he received information about the postponement of the congress in December 1936. For instance, he had not gone to Moscow because his own suggestion to visit had not been answered for months, and when the Russians urged him to come, he had already "collected quite some information." He felt that under the circumstances, a visit would not allow him to speak freely with the Russian geneticists, while it might bind his actions. Reading the bulletins from the December 1936 congress had made "a very strong impression," Mohr told Dunn. They showed "how Lysenko's followers could—for party reasons—permit themselves of brazen attacks on well-founded genetic facts," and how "Serebrovskii in a most humiliating way had to withdraw opinions which he had expressed years before." And they contained astounding statements like "the only scientific method is Marx's,

Lenin's, Stalin's method." Mohr was careful not to give such arguments explicitly in his official correspondence. But it is clear that for Mohr, the theory and practice of Stalinist science policy that could now be observed in the Soviet Union were sufficient reasons for rejecting the thought of a Moscow congress. Already in January, Mohr had confided to Dunn that even if the rumors about arrests and so on were exaggerated, "I think we may as well immediately face the situation that cancellation and not only postponement will be the final result." But he agreed with Dunn that "everything should be done to prevent a break" that could hurt international scientific cooperation or the Russian geneticists in their difficult situation.[78]

But what about the responsibility for Soviet colleagues? "Ought we to go to Moscow in order to give support to the decent geneticists who are in a critical situation?" asked Mohr in his letter to Dunn. After discussing the point with his old friend the Norwegian ambassador to Moscow, Mohr had concluded that "open support to them may as well bring them into greater difficulties."[79] A month later Mohr wrote to Dunn that he had spoken to a British scientist—a neurologist interested in genetics—who had just been to Russia. He had said that it had been impossible to talk to Gershenson, Levit, and others without an "interpreter" being present all the time. "Suspicion everywhere, and all conversations started with cliché declarations about the Marxist science being the only science etc., etc." Mohr had shown him the letter of June 22 signed by all the Russian geneticists. "He said you couldn't trust it at all, they simply *had* to."[80]

In November 1937 Muller, who had by then definitively left Moscow, wrote to Mohr that the Soviet geneticists now shared his judgment:

> Confidentially, I would say it was not a mistake, the way the internatl. conf. was finally decided. For altho the Russian geneticists really did want the congress there still, at the time I wrote to you urging it, conditions had changed enough by this September that they had changed their minds already, and were hoping the decision would be the other way, as not to create a still more embarrassing situation for them.[81]

The whole affair had been a strong personal strain on Mohr, and he was very relieved to receive this letter from Muller. The whole atmosphere of secrecy, lies, and distrust was disgusting to such an apostle of science, truth, and liberty as Mohr:

> Without any comparison this matter has been the most difficult to deal with in my experience. Not to be able to trust the information you receive creates a feeling that is unbearable. And not to dare write openly, even to friends, for fear that the letter may bring them in difficulties, feels still worse. This has been my situation versus you for instance.[82]

Three months later Muller wrote a letter of consolation to Vavilov: the representatives of the "two great English-speaking countries (Haldane for England and Dunn for America)" had voted in favor of Moscow. Thus the decision was taken against the votes of the two "most important countries in genetics outside the USSR itself." Muller indicated that he himself had been an active and influential lobbyist: "Perhaps I flatter myself, but I think my efforts had something to do with the way these two countries reacted."[83] Muller thus confirmed the strain and difficulty of Mohr's political balancing act.

One purpose of telling this detailed story about the cancellation of the international genetics congress in Moscow is to give a picture of how the international scientific network was able to function even under difficult political circumstances. We see how important the personal relationship between the leading scientists and their direct responsibility for the decision-making process was. A more bureaucratic international organization would have great difficulty in seeing its way through this maze of political terror, pressure, and propaganda, of rumors, half-truths, and concealments.

Another purpose of this account of the cancellation of the Moscow genetics congress is to show in detail how O. L. Mohr's course of action was guided by his classical liberal ideals of science. Despite censorship and terror in the Soviet Union, Mohr was able to form a pretty accurate assessment of the situation, an assessment that in the end agreed with that of the Soviet geneticists themselves. Political powers interfered with scientific autonomy in a way that the international scientific community would not accept. While extramaterial support for dubious scientists such as Lysenko, due to political sympathy, may be a waste of public resources, it would not be incompatible with the liberal ideal of science. It would be part of the noise one has to live with in an open democratic society. But the Soviet attempts to set the agenda for scientific discussions and censor scientific arguments and opinions was not acceptable. When knowledgeable Western scientists, such as J. B. S. Haldane, John Desmond Bernal, and Joseph Needham during the 1930s and 1940s, defended Lysenko's activities in genetics as being legitimate science, the recurrent argument was that he represented one side in an open scientific debate. As long as his opponents were free to continue their work and express their opinions, Lysenko's activity had to be accepted even if it was in some respects pseudoscientific. When Julian Huxley condemned Lysenkoism in his 1949 book *Soviet Genetics and World Science*, it was the closure of open scientific debate that was his central argument against Lysenkoism, not the falsehood of Lysenko's claims.

Could a 1938 international congress in Moscow have helped the Soviet geneticists in their struggle against Lysenkoism? We have seen that by the autumn of 1937 Muller reported that even the Soviet geneticists themselves agreed that it could not. Vavilov did function as president of VASKhNIL in late 1937 and early 1938. Apparently, he had considerable support in the party apparatus, and it is imaginable that clever diplomacy could have given him the presidency again instead of Lysenko. But it seems unlikely that an international congress could have supported the kind of subtle lobbying that was needed. Under the circumstances, it would be more likely to provoke contrary political forces by its pressure for openness and freedom in scientific debates, including liberty to discuss eugenics and race theories.

NOTES

1. For the article by Lysenko and Prezent see *Iarovizatsiia* 3 (1937): 49–66.

2. For an engaging account of Kol'tsov's fight to preserve scientific freedom in the late 1930s, see Gaissinovich and Rossianov (1989).

3. Kol'tsov's letter and most of Muralov's reply were published by Gaissinovich and Rossianov (1989).

4. Quotes taken from Reznik (1983, 77).

5. TsGAE, f. 8390, o. 1, ed.khr. 954, l.81–82.

6. TsGAE, f. 8390, o. 1, ed.khr. 955, l.36

7. After a broad "democratic" process of discussion through 1936 the new constitution was enacted by the Congress of Soviets on November 27. See, for instance, Tucker (1992, 354).

8. TsGAE, f. 8390, o. 1, ed.khr. 953, l.46

9. TsGAE, f. 8390, o. 1, ed.khr. 955, l.55.

10. TsGAE, f. 8390, o. 1, ed.khr. 954, l.79.

11. TsGAE, f. 8390, o. 1, ed.khr. 953, l.48–50.

12. TsGAE, f. 8390, o. 1, ed.khr. 954, l.75–76.

13. TsGAE, f. 8390, o. 1, ed.khr. 954, l.81.

14. TsGAE, f. 8390, o. 1, ed.khr. 954, l.82–84.

15. TsGAE, f. 8390, o. 1, ed.khr. 954, l.30.

16. TsGAE, f. 8390, o. 1, ed.khr. 955, l.37–39.

17. TsGAE, f. 8390, o. 1, ed.khr. 955, l.40.

18. TsGAE, f. 8390, o. 1, ed.khr. 955, l.31.

19. TsGAE, f. 8390, o. 1, ed.khr. 955, l.71.

20. TsGAE, f. 8390, o. 1, ed.khr. 955, l.48–49.

21. TsGAE, f. 8390, o. 1, ed.khr. 955, l.55, 66–69.

22. TsGAE, f. 8390, o. 1, ed.khr. 953, l.33–43.

23. TsGAE, f. 8390, o. 1, ed.khr. 956, l.42–48.

24. TsGAE, f. 8390, o. 1, ed.khr. 954, l.97–100.

25. TsGAE, f. 8390, o. 1, ed.khr. 956, l.84–86.

26. TsGAE, f. 8390, o. 1, ed.khr. 955, l.62–63.

27. The intravarietal cross-fertilization was to be effected by cutting off the stamens with scissors and thus preventing self-fertilization. Lysenko's idea was that this should be done for all seed production.

28. TsGAE, f. 8390, o. 1, ed.khr. 955, l.113–14.

29. TsGAE, f. 8390, o. 1, ed.khr. 956, l.57–58.

30. TsGAE, f. 8390, o. 1, ed.khr. 956, l.81–82.

31. TsGAE, f. 8390, o. 1, ed.khr. 955, l.57.

32. TsGAE, f. 8390, o. 1, ed.khr. 955, l.27–43.

33. Nikolai Krementsov has done a thorough study of the process that led to cancellation of the 1937 genetics congress in Moscow. In Doel et al.

(2003), he presents a detailed and well-documented account of the negotiations between the scientists and the political powers, pointing to ideological as well as practical reasons that led to the cancellation by the Politburo in November 1937 and to the postponement to August 1938 obtained in the spring of 1937. His description brings out the strong political will to censor the agenda of the congress. He also cites a number of Western geneticists worrying about the similarity of the Soviet Union and Nazi Germany in suppressing scientific freedom. These topics will also be covered extensively in Krementsov's book *Science between the World Wars: The Case of Genetics*, announced to be published by Routledge in November 2004.

34. Letter from N. I. Vavilov to Ia. A. Iakovlev, March 23, 1932, *VL* 2, pp. 163–64.

35. Among those who did not turn up were G. D. Karpechenko, M. S. Navashin, G. A. Lewitskii, A. S. Serebrovskii, N. P. Dubinin, and V. Pisarev. The only prominent Soviet geneticist besides Vavilov that did register at Ithaca was A. A. Sapegin. Altogether three Russians registered. (*Proceedings of the Sixth International Congress of Genetics, Ithaca, New York, 1932*, vol. 1, edited by D. F. Jones [New York: Brooklyn Botanic Garden, 1933], p. 17.)

36. Letter from H. J. Muller to O. L. Mohr, April 22, 1935, Archive of the Institute for Medical Genetics, University of Oslo; letter from N. I. Vavilov to O. L. Mohr, June 3, 1935, Muller Archive, Lilly Library, University of Indiana, Bloomington.

37. Letter from N. I. Vavilov to O. L. Mohr, August 29, 1935, Muller Archive.

38. Letter from H. J. Muller to O. L. Mohr, April 22, 1935. Archive of the Institute for Medical Genetics.

39. Letter (circular) from O. L. Mohr to H. F. Federley, August 29, 1935, Federley papers, University Library, Helsinki.

40. Letter (circular) from O. L. Mohr to H. F. Federley, November 14, 1935, Federley papers.

41. Letter from N. I. Vavilov to O. L. Mohr, December 7, 1935, Muller Archive.

42. Decision of SNK, February 3, 1936, ARAN, f. 201, o. 3, d. 2, l.3.

43. For Levit's biography see Adams (1990b).

44. Meeting of March 17, 1936, ARAN, f. 201, o. 3, d. 54, l.7–8.

45. Report from a meeting in the program committee for the Seventh International Congress of Genetics on April 5, 1936, ARAN, f. 201, o. 3, d. 54, l.4.

46. Protocol from meeting of organizing committe of Seventh Interna-

tional Genetics Congress, April 23, 1936, appendix no. 1, ARAN, f. 201, o. 3, d. 3, l.23.

47. Note on the "program of the Seventh International Congress of Genetics," containing "unanimous decisions" of the organizing committee. Signed by Muralov and Levit, September 26, 1936. ARAN, f. 201, o. 3, d. 3, l.1–5.

48. This could hardly have been anybody else than Eugen Fischer, director of the Kaiser Wilhelm Institute for Anthropology, Human Genetics, and Eugenics in Berlin.

49. Letter from H. J. Muller to O. L. Mohr, October 23, 1936, Archive of the Institute for Medical Genetics.

50. ARAN, f. 201, o. 3, d. 10, l.1–3. This memo is dated October 8. Head of the three-person commission was Shaxel.

51. ARAN, f. 2, o. 1–1935, d. 83, l.101. The letter is dated November 16, marked "personal" (*lichno*), and signed Prokhorov.

52. ARAN, f. 2, o. 1–1935, d. 83, l.104. This copy has no precise date, only "November." It is signed by Krzhizhanovskii (vice president) and Gorbunov (permanent secretary).

53. Letter from H. J. Muller to O. L. Mohr, dated Moscow, December 8, 1936, Archive of the Institute for Medical Genetics.

54. Letter from O. L. Mohr to H. Federley, April 8, 1937. This was a circular apparently sent to all members of the Permanent International Genetics Committee. With this circular was enclosed a copy of a letter from O. L. Mohr to A. I. Muralov of January 7, 1937. This latter letter also contains information on the exchange of telegrams and the reaction to the notice in the *New York Times*. (Federley papers.)

55. ARAN, f. 201, o. 5, d. 3, l.11.

56. Letter from Mohr to Dunn, May 28, 1936, American Philosophical Society Library, Philadelphia, Pennsylvania.

57. Letter from Mohr to Dunn, December 6, 1936, American Philosophical Society Library.

58. Letter from Mohr to Muralov, January 7, 1937, Federley papers.

59. Letter from Mohr to Dunn, January 25, 1937, American Philosophical Society Library.

60. Letter from Muralov and Vavilov to Mohr, February 13, 1937, Fererly papers. Transcript sent by Mohr to the members of the International Genetics Committee.

61. Letter from Mohr to Muralov, March 13, 1937, Federley papers. Transcript circulated by Mohr.

62. ARAN, f. 201, o. 3, d.16, l.8. Protocol from meeting of the Komarov commission on March 25, 1937. The telegram was signed by Komarov as president of the Academy of Sciences, and Muralov and Vavilov on behalf of the organizing committee. Present were also Gorbunov, Gershenson, Sapegin, Levit, Lysenko, Kostov, Bankin, and Serebrovskii.

63. Letter from Mohr to Dunn, August 4, 1937, American Philosophical Society Library.

64. Memorandum from Mohr to the members of the Permanent International Committee of Genetics, April 8, 1937, Federley papers.

65. Then foreign minster of the Soviet Union.

66. Vavilov to Muller, April 5, 1937, Muller Archives. Quoted from Carlson (1981, 238–39).

67. Letter from Vavilov to Federley, April 5, 1937, Federley papers. In German: "Da die Frage des Rassenproblems bei uns überhaupt nicht existiert, so hat sich auch unsere Regierung gegen ihre Aufnahme in das Programm des Kongresses ausgesprochen."

68. ARAN, f. 2, o. 1–1935, d. 83, l.57–57ob. Protocol from meeting of the presidium of the Academy of Sciences, May 5, 1937.

69. ARAN, f. 2, o.1–1935, d. 83, l.52.

70. Letter from Muller to Mohr, May 18, 1937, Archive of the Institute for Medical Genetics.

71. Letter from Mohr to the organizing committe, in the name of Muralov and Vavilov, May 20, 1937, ARAN, f. 201, o. 5, d. 4, l.28–29.

72. Letter from Vavilov to Mohr, May 23, 1937, ARAN, f. 201, o. 5, d. 4, l.26.

73. Meeting of the organizing committee, June 16, 1937, ARAN, f. 201, o. 3, d. 3, l.8–9.

74. Letter from Komarov, Muralov, and Vavilov to Mohr, June 17, Federley papers. Transcript sent by Mohr.

75. Letter to Mohr, June 22, 1937. See memorandum from Mohr to the International Committee, July 21, 1937, Federley papers. The signatories were G. K. Meister, N. I. Vavilov, M. S. Navashin, N. K. Koltzoff, A. S. Serebrovskii, D. Kostov, S. G. Levit, N. P. Dubinin, A. A. Sapegin, D. A. Kislovskii, S. M. Gershenson, G. A. Levitskii, and G. D. Karpechenko.

76. Memorandum from Mohr to members of the Permanent International Committe on Genetics, July 21, 1937, Federley papers.

77. Letter from Mohr to Komarov and Meister, September 23, 1937, and letter from Mohr to Federley, same date, Federley papers. The letter to the Moscow organizing committee was addressed to Komarov as president

and Meister as secretary. Mohr apparently knew that Muralov had now been arrested.

78. Letter from Mohr to Dunn, January 25, 1937, American Philosophical Society Library.

79. Letter from Mohr to Dunn, August 4, 1937, American Philosophical Society Library.

80. Letter from Mohr to Dunn, September 9, 1937, American Philosophical Society Library.

81. Letter from Muller to Mohr, November 22, 1937, Archive of the Institute for Medical Genetics.

82. Letter from Mohr to Muller, November 26, 1937, Muller Archives.

83. Letter from Muller to Vavilov, February 21, 1938. ARAN, f. 2, o. 1–1935, d. 83, l.32.

Chapter 9.

Lysenko Takes Over

With the events of 1936–1937, the stage was set for Lysenko's takeover of Soviet agricultural science, and subsequently genetics. This development is perhaps most clearly illustrated by the fate of the International Congress of Genetics. Otto Mohr, as chairman of the international committee, viewed the events from a distance with insufficient details of information. But by early 1937 he saw quite clearly that if the Soviet government did not change its politics of science, and the policy toward genetics in particular, there was no hope for an acceptable international congress—and Soviet genetics would enter hard times.

From its start in 1929 the Lenin Academy of Agricultural Science (VASKhNIL) had been under strong political control. This control was primarily economically motivated. In order to get the most use out of scientific research, it had to be consciously directed toward the tasks that were most important for the agricultural economy. The scientists themselves could not be trusted with this. They were too likely to get bogged down with problems they found scientifically important and interesting. As time went on, the demand for efficiency became more pronounced. Both time and resources were scarce, and the political leadership wanted concentration on the research that gave the most profit relative to the resources invested. Demand for greater economic efficiency was the driving force in the

1934–1935 reorganization of VASKhNIL. Vavilov as president had not given this aim sufficient priority.

It was the criterion of economic profit that put Lysenko in the forefront of agricultural research. In the early 1930s the phenomena of vernalization, the effects of temperature and light on the development of plants, were new. And it was quite reasonable to hope that important new agricultural techniques might be developed on their basis.

But it was the demand for political loyalty that propelled Lysenko to the top. We saw in the preceding chapter how political loyalty took priority in the genetics controversy in 1936–1937. For a while *partiinost* (party-mindedness) in part displaced practical results as a criterion for evaluating agricultural research. This was a main component in the tightening of political control in VASKhNIL in 1936–1937. The strong demand for party-mindedness in science also made eugenics and race theories relevant to genetic theory when they were pursued in the context of plant and animal breeding. Prezent saw this clearly and made good use of the new situation drawing on his experience from the cultural revolution at the beginning of the 1930s.

LYSENKO BECOMES PRESIDENT OF VASKhNIL

By 1936–1937 Lysenko had become a major public symbol of the new proletarian science. This, combined with his standing as a vice president of VASKhNIL, made him a significant candidate for the top position in Soviet agricultural research. He had formal scientific legitimacy as well as the necessary political support. However, in early 1937 there were still big obstacles to Lysenko's rise to the very top. Three scientists in the VASKhNIL presidium, Vavilov, M. Zavadovskii, and Meister, clearly opposed Lysenko's ideas on breeding and genetics and would do what they could to prevent him from becoming president. While the first two were ideological outsiders, Meister was a party member and ardent prophet of the new Soviet type of science. We have also seen that the political leadership of VASKhNIL, in particular the president, Muralov, had doubts about the soundness of some of Lysenko's ideas. Working in VASKhNIL for a while, Muralov was aware of the lack of hard evidence for Lysenko's claims of superior practical methods. But because of Lysenko's popular fame and political support, they all tended to appease him as best they could.

The great terror helped Lysenko over the last hurdle. Muralov, Margolin, and Meister, the very troika that in January 1937 had rejected Kol'tsov's appeal on behalf of science, were all arrested in the summer of 1937 (Gaissinovich and Rossianov 1989, 96–97). The arrests probably had little to do with the dispute over genetics. They were primarily a result of the internal power struggle in the party, ending with the liquidation of the old guard of Bolsheviks and independently thinking party members. Another important VASKhNIL academician arrested and suppressed during the same period was Tulaikov. He had been heavily criticized for his agrotechnical method of "shallow plowing" by Muralov and Bondarenko during the VASKhNIL *aktiv* in March. All four—Muralov, Margolin, Meister, and Tulaikov—were Communist Party members. Their disappearance opened Lysenko's access to the very top political and administrative position in agricultural science.

Vavilov served as acting (*ispolniaiushchii obiazannosti*) president of VASKhNIL for a while until Lysenko was named president in February 1938. On April 1, 1938, the presidium of VASKhNIL met for the first time under the presidency of Lysenko.[1] Vavilov and M. Zavadovskii remained vice presidents, while Nikolai V. Tsitsin took Meister's place as the third vice president. Once more chance had smiled at Lysenko. Not burdened by an upper- or middle-class background and not involved in party politics, he ran a relatively low risk of being repressed. But it was not pure chance. We have seen how powerful political and ideological trends were in shaping science in a way that favored Lysenko, with respect to research problems, ideology of method, and organizational principles.

As president of VASKhNIL Lysenko occupied the top position in the hierarchy of agricultural science. He also represented science in various political organs of the Soviet state. For instance, he became a member of the Supreme Soviet (the Soviet parliament) in 1938, as

Vavilov had been earlier. Lysenko even had the honor of being a vice president in one of its two chambers.

Lysenko was now Vavilov's superior. He could in various ways intervene in the work of VIR, Vavilov's Institute for Plant Industry, supporting people who shared his own opinions and obstructing the interests of Vavilov. For instance, he excluded Zhebrak and Karpechenko from participation at the national agricultural exhibition, and he intervened to include persons of his own views in the scientific council of VIR.[2] But Lysenko did not have the power directly to discharge his most prominent opponents from their posts. He could not remove Vavilov as leader of VIR without the approval of the Central Committee, and he had to put up with M. Zavadovskii and Vavilov as vice presidents of VASKhNIL. Though Lysenko could generally count on the support of Tsitsin, VASKhNIL was more or less immobilized by the split in its presidium and other opposition within the organization. This conflict affected important practical questions of seed growing, and plant and animal breeding, and was a continuing headache for the Ministry of Agriculture.

However, Lysenko gradually increased his administrative power in Soviet science. He and his supporters put effective political and ideological pressure on opponents in policy debates. We have seen that the December 1936 conference organized by VASKhNIL was a big success for Lysenko. It strengthened the broad sympathy for his views in the political establishment and the mass media and changed the political climate within the research institutions in his favor. This success continued with the conference on genetics and breeding in October 1939. And it culminated with the notorious 1948 VASKhNIL conference called "The Situation in Biology."

ROLE OF PHILOSOPHY—1939 CONFERENCE

Western scientists followed the developments in Soviet genetics with anxiety. The American *Journal of Heredity* reported on a meeting of 150 breeders and geneticists held at the Ministry of Agriculture in March 1939, which signaled the effects of Lysenko's takeover of the hegemony in agricultural research. Prezent ridiculed Mendelism. Lysenko urged liberation from the Mendelism that "still sits within every one of us." Eikhfel'd claimed that observation of nonmanipulated nature showed that Lysenko was right. The head of elite seed production in the Soviet Union declared that future work in plant breeding and seed production would be based on "Darwinism and Michurinism," and the minister of agriculture supported this view. Only a minority headed by Vavilov and Zhebrak defended classical genetics as the basis for plant breeding and seed production, according to this report.[3]

To improve their position, the geneticists wanted a new public discussion on genetics after the setback following the December 1936 congress. They hoped to disseminate knowledge about the scientific weaknesses of the Lysenkoist views and to make the Communist Party and the government aware of the way in which Lysenko was misusing his administrative power to suppress classical genetics. A letter from a group of Leningrad biologists to Andrei Zhdanov, head of the *Agitprop* (agitation and propaganda) division of the Central Committee, was instrumental in bringing this new conference about.[4] Zhdanov was apparently impressed by complaints about the "adminstrative methods" used by Lysenkoists. He also noted objections about the lack of factual foundation for Lysenko's genetic theories and their anti-Darwinian character. At the end of June 1939, the secretariat of the Central Committee decided that the editorial board of the party's theoretical journal, *Under the Banner of Marxism*, was to organize a new conference on the disputed issues of genetics.

This conference was held in Moscow on October 7-14, 1939.[5] Philosophers from the editorial board of the journal, including Mark B. Mitin, P. F. Iudin, and Arnost Kol'man, played a central role at the conference as chairmen and speakers. After the conference the philosophers produced a report that was sent to the Central Committee. At the December 1936 conference the main report and evaluation was given by a specialist in the scientific disciplines concerned, namely, Meister. In 1939 it was given by a philosopher, Mitin, head of the Institute for Philosophy of the Academy of Sciences. This reflected and reinforced the new situation. Philosophical and ideological arguments had become more important. The philosophers acted as a priesthood, arbitrating a scientific controversy in the name of society. While the older generation of scientists, educated before the 1917 revolution, viewed this role with skepticism and suspicion, it was apparently accepted as natural by the younger generation steeped in Marxism. They tended to become evasive when pressed on questions where genetic science bordered on the teachings of dialectical materialism.

Mitin's brief introductory talk pointed to the need for ideological reeducation of scientific cadres. Since there was "no Chinese wall between theory and practice in the land of socialism," a correct resolution of the theoretical scientific questions had immense practical importance, he explained. Michurin was a brilliant example of the achievements of science under the new and favorable conditions for "progressive science" created by socialism. The most important task of scientists was to study and master dialectical materialism. Many Soviet scientists still "accept uncritically . . . reactionary theory and obsolete approaches."[6] Mitin made it quite clear what was the yardstick of progressive science.

Vavilov was now very explicit and clear in his criticism of Lysenko's genetic ideas. There was no more compromising or benevolent accommodation. He stressed that the great achievements of modern plant breeding, in the Soviet Union as well as in other countries, had been based on the distinction between genotype and phenotype introduced by Johannsen. The "Svalöf selection station has been accepted by the common consent of geneticists and selectionists as the leading selection institution in the world for both theoretical and practical work." Lysenko had rejected this basis. Matters had come so far that the Ministry of Agriculture had decided to follow his advice and "change radically the methods of the selection stations." But there were "no experimental data whatever" to show the need to depart from the established methods, explained Vavilov (1939, 187–88).

Vavilov's defense of the chromosome theory was equally clear: "To deny the role of the chromosomes, to attempt to explain everything in terms of the organism as a whole or of the cell, is to set biological science back a century." He also condemned the substitution of vegetative hybridization for sexual hybridization in plant breeding that was taking place on the initiative of Lysenko. It lacked a sound scientific basis and was causing great harm to Soviet agriculture (Vavilov 1939, 191–92).

Lysenko concentrated his attack on theoretical issues of genetics, in particular the law of segregation, that is, the 3:1 ratio in the second generation of offspring, which had been the trademark of Mendelism since its inception.[7] This ratio became the focus of the discussion. Had it been proved to be a real fact of nature or not? Lysenko interpreted the Mendelians to hold that this was a strict, universally valid empirical law:

> ... the offspring of any hybrid (just think of it: any hybrid!) of all plants and all animals must of necessity vary according to one and the same pattern, independent of the variety and genus of the animal or plant, of the conditions of life, or of any other possible influences. Always and everywhere it will be $(3:1)n$. (Lysenko 1939, 203–204)

But, of course, the 3:1 ratio was not meant to be a universal empirical law in this strict sense. First, it was a statistical and not a deterministic law. And second, exceptions to the statistical law were everywhere and well known. What classical genetics claimed to have discovered was a cytological mechanism, namely, the existence of paired hereditary factors (genes) lodged in paired chromosomes. Under favorable circumstances, that is, when other causal factors did not intervene, this mechanism produced the 3:1 ratio. The regular occurrence of this ratio, or readily explainable modifications of it, in controlled experiments, was strong evidence for the existence of such a mechanism of biological inheritance.

Lysenko rejected the cytological theory as speculative and metaphysical: "this regularity is statistical, not biological." At the 1939 conference this line of argument was energetically pursued by Kol'man, philosopher and mathematician. For instance, he broke into the talk of the young geneticist Iu. Ia. Kerkis, arguing that a classification of flies into black and gray was much too arbitrary and subjective to prove a 3:1 ratio. Prezent followed up by asking the speaker if his suit was gray or black.[8]

Later in the debate Kol'man explained his objection with an analogy. Even for a batch of buns fresh from the oven, you can easily establish a 3:1 ratio between small and large. It just depends on where the line is drawn.[9] Kol'man with appreciation quoted Karl Pearson's *Grammar of Science*: "Scientific laws are more a product of the human mind than a fact of the external world."[10] He also referred to the "physiological genetics" of the German American geneticist Richard Goldschmidt, who had rejected the idea of genes as separable material entities (corpuscles) in favor of genes as interpenetrating "structures."[11] This empiricist-inspired rejection of Mendelian genetics had something in common with Pearson's way of brushing off Johannsen's genotype theory and pure lines back in 1903. The dislike of metaphysics and nonobservable entities was pushed so far that it disqualified crucial theoretical entities in science, such as genes (Roll-Hansen 1989). The radical empiricist heritage from Ernst Mach through Bogdanov was alive and active.

Throughout the conference, philosophers interrupted to ask geneticists about their philosophical views and their opinions on Michurin. Philosophical "idealism" and skepticism about the achievements of Michurin were mistakes that needed to be revealed. For instance, after Kerkis had finished his talk, Mitin asked about his "attitude toward the theoretical works and the practical achievements of Michurin." The "practical results of Michurin are beyond doubt," answered Kerkis. "And the theoretical?" retorted Mitin. Kerkis went on to explain at some length: "I hear laughter, but I repeat once more that no doubt about the practical importance of the theoretical results of Michurin has arisen in my mind. Concerning the theoretical explanations, I reckon that in some respects Michurin was mistaken, in particular I consider the theory of mentors[12] mistaken in many ways." Kerkis hastened to add that though he did not know all the practical results of Lysenko's work well, he did not doubt them. But his words nevertheless indicated underlying doubt: "apparently vernalization is in many cases, but not always, as T. D. has thought and written."[13]

Kerkis had presented his own critical repetition of some Lysenkoist experiments claiming to disprove the 3:1 Mendelian ratio. Throughout the talk he was heckled, especially by Lysenko. An exchange at the end demonstrates the grumbling nature of Lysenko's objections. Kerkis had just remarked that insulting or abusive characterizations such as "anti-Darwinism," "racism," and "pseudo-science" were as harmful to the "the school of Trofim Denisovich [Lysenko]" as to the geneticists.

> Lysenko: I don't have a school.
>
> Kerkis: Well, then to the group of Trofim Denisovich.
>
> Lysenko: I do not have a group, I am a Michurinist.
>
> Kerkis: In that case for the direction presently headed by Trofim Denisovich.
>
> Lysenko: I am president of the academy.[14]

Zhebrak in his talk complained briefly about the difficulties geneticists met, for instance, when they wanted to correct the lies that their critics published in newspapers. But he quickly went on to substantial scientific issues emphasizing that dialectical materialism was the theoretical framework for his science of genetics. His main thesis was that classical genetics presents "a confirmation of the laws of dialectics and primarily the law of the transformation of quantity into quality."[15]

Zhebrak, like Vavilov, was uncompromising in his defense of Johannsen's pure line theory as the basis for the success of modern plant selection. But he used stronger and more direct language. The main results of Michurin, said Zhebrak, were based on hybridization in full accordance with classical genetics. There was nothing in his results to contradict Mendelian genetics. Lysenko had also used the ordinary methods of hybridization, according to Zhebrak. He did not have any special method. And he had not produced a new variety of wheat in two and a half years, as he claimed. Now five and a half years had passed and Lysenko's variety was still of poor quality. In conclusion, Zhebrak proposed that a commission should be established to present a "fully objective verification of the facts" of the debate, independent of any specific point of view or group attachment. The philosopher chairman, however, reacted sharply at his suggestion of members for the commission: "But they are enemies of the people." And from the audience it was remarked that Zhebrak wanted "politically to disgrace Lysenko."[16] According to the official summary of the conference, Zhebrak's talk "suffered from a complete lack of self-criticism" (Kolbanovskii 1939, 99).

Another young, uncompromising, and philosophizing Marxist geneticist and Communist Party member was Nikolai P. Dubinin. He started by expressing his respect for Lysenko as a great scientist and the president of VASKhNIL but quickly continued with severe criticism. Dubinin concentrated on defending the Mendelian 3:1 ratio as a fact of nature. He criticized Lysenko's claims about genetic change through "reeducation," "vegetative hybridization," method of "mentors," and so on.[17] At the same time he claimed to be a true Michurinist. Vernalization as a physiological method he did not discuss. Dubinin concluded by declaring that in spite of disagreements, they all had the same basic platform, "class science" and "service for the people."[18]

Serebrovskii was glad the conference had been organized. Soviet geneticists had experienced many difficulties lately. He saw the conference as a chance to defend his scientific views and correct the public perception of classical genetics. In conclusion, Serebrovskii asked the editorial board of *Under the Banner of Marxism* to help secure a reliable and accessible overview of the factual basis of genetics. This would promote fruitful investigation of theoretical problems as well as practical economic usefulness. Iudin followed up by asking who, in the view of Serebrovskii, represented idealism in Soviet genetics. He responded hesitatingly, mentioning Filipchenko's autogenetic ideas and Lev S. Berg's similarly anti-Darwinian views. When Iudin asked for the third time whether idealist views still occurred in Soviet genetics, the answer was no. Idealists are those who "do not understand the leading role of natural selection in evolution," declared Serebrovskii.[19] But this was not a satisfactory answer in the eyes of his interrogator. It revealed that Serebrovskii was unable to see or admit his own idealist tendencies. To the philosophers assertion of unproven "metaphysical" theories, like Mendelism, was idealism.

The brothers Zavadovskii took contrary positions. Mikhail Zavadovskii continued his valiant defense of classical genetics. Boris Zavadovskii, on the other hand, disliked the polarization of the debate and claimed to hold a "third point of view." However, his view was also that the main problem was the idealist philosophical position of the Mendeliams. Lysenko was "95–96 percent right."[20]

Lysenko turned the practice criterion against Vavilov. He quoted some bombastic claims that Vavilov had made in the early 1930s about the divorce of plant breeding from genetics as an example of the alienation between theory and practice in traditional bourgeois science. With some right Lysenko held these claims to be contrary to Vavilov's present defense of Mendelism because of its practical usefulness. But to say that Vavilov and Serebrovsii had then declared "that Mendelian genetics has no relation whatsoever to the derivation either of varieties of plants or breeds of cattle" was clearly an exaggeration (Lysenko 1939, 207–209). Lysenko made it appear as if he was now defending Vavilov's revolutionary position of 1932, while Vavilov had defected to the enemy camp. In Lysenko's interpretation of the events, it was he who had presented the Soviet public and government with a new genetic theory. This theory had already achieved considerable practical successes, while the lacking fruitfulness of classical genetics was in fact being confirmed by classical geneticists themselves, he argued.

Lysenko's argument fitted well with the philosophers' ideas about the role of the practice criterion in the evaluation of science. Their lack of biological and agricultural knowledge made them easy victims of wishful thinking. According to the organizers, the conference had given particular attention to "the practical achievements" of Lysenko and his collaborators.[21] But as we have seen, it was mainly scientific experiments relevant to Mendelian theory that were discussed. Doubts about the economic efficiency of Lysenko's practical methods, which Vavilov, Zhebrak, and some others raised, were neglected by both the philosophers and the Lysenko supporters. The philosophers also seemed to forget that the success or failure of a scientific experiment is something very different from the economic efficiency or inefficiency of a practical method based on it. The vernalization of winter grains was perfectly feasible, but it was not economically profitable. The philosophers' charge of a fatal split between theory and practice was ironically accurate. It applied to themselves rather than the geneticists. Instead of subjecting the practical usefulness of Lysenko's methods to thorough scrutiny, they put most of their effort into a methodological discussion of theoretical genetics. As one of Vavilov's colleagues from VIR remarked, "the leading selectionists of the country" were not even present.[22] That scientific methodology was the only aspect of the genetics debate where philosophers had some competence and could make meaningful contributions underlines the irony.

Mitin gave the authoritative summing up of the congress. His conclusions implied mild criticism of Lysenko's "progressive" position and a serious warning to the "conservative" geneticists. Lysenko needed to take the positive facts produced by classical genetics more seriously, for instance, the knowledge about chromosomes. Also, some of his supporters, such as Prezent and Grigorii Shlykov, a young radical and activist on the VIR scientific staff (Joravsky 1970, 119; Popovskii 1984, 140–41), were sometimes failing to observe the rules of proper scholarly conduct. But the geneticists were committing a more fundamental and serious sin. They were tenaciously adhering to mistaken methodological principles.

Timiriazev and Michurin were right, stated Mitin. Mendelism was valid for a number of special cases, but not as a general law. The metaphysical tendency of Mendelians to universalize their laws conflicted with dialectical materialism. Mitin repeated the claim that the concept of the "gene" prevalent in classical Mendelian genetics was inconsistent with the theory of evolution (Mitin 1939, 160–65). One textbook in particular was attacked for "solid metaphysics" in its treatment of the genotype/phenotype distinction (Mitin 1939, 170).[23]

The association of Mendelian genetics with eugenics, racism, and the like was mentioned by both its critics and its defenders. But accusations of this kind played a relatively minor role in the criticism. Kol'man attacked Vavilov and his collaborator, Rozanova, for a passage in a book by Johannsen describing the study of human races as a prime task of genetics. They had edited the book and written the preface, but had not warned the readers, said Kol'man. It was revealed during the talk that the passage came from another book by Johannsen translated into Russian a few years earlier. But Kol'man still maintained that they should have criticized this eugenic view held by Johannsen.[24]

For Mitin the great practical achievements of Lysenko were beyond doubt. "Against this nobody has said, or could say, anything from the rostrum of our meeting, because these things have been introduced into practice, into life, and have been widely disseminated" (Mitin 1939, 150). The criticisms that had been raised clearly made little impression on him or the other philosophers. However, it must be taken into account that even among the geneticists opposing Lysenko, there was still a tendency to take his practical achievements outside of genetics for granted. Two of the most insistent and knowledgeable critics of Lysenko's practical agricultural achievements, Lisitsyn and Konstantinov, did not participate in the discussions. To prove that Lysenko had a high reputation for "practical and theoretical achievements" in the West as well as in the Soviet Union, Mitin (1939, 151) referred to a review article by R. O. Whyte, "Phasic Development of Plants" (Whyte 1939). The noncommittal descriptive reporting of the Imperial Bureaus of Plant Breeding was important for Lysenko's scientific prestige in the Soviet Union.

The practice criterion of truth was the central principle in the philosophers' evaluation of genetics in 1939. They took Lysenko's practical success for granted and showed how his methodological and theoretical claims fitted with dialectical materialism. It had become a dogma that Lysenko had succeeded in unifying theory and practice, while traditional genetics suffered from their separation. As I have mentioned, it was in the 1930s a common judgment among plant breeders also in the West that breeding was suffering from a separation of theory and practice. Classical genetics had not yet delivered the fruits that were promised in the early years of the twentieth century (Roll-Hansen 2000).

A comparison of the 1936 and 1939 conferences shows how the balance of power had been shifted in favor of Lysenko. In 1939 the philosophers generally sanctioned the Lysenkoists' methodological and ideological criticism of classical genetics. The room for scientific argument was narrowing as philosophical orthodoxy tightened its grip on the correct interpretation and application of scientific method. In 1939 the philosophers acted as umpires, crying foul whenever they perceived a violation of Marxist philosophy or the canonized texts of Soviet science. They did not perceive the ironies of their own activity. The effect of their evaluation of biological theories in the name of practice was to make them less useful for Soviet agriculture. The philosophers lacked sufficient knowledge of agricultural science and practice to understand what they were doing. It is not surprising that the same applied to their political bosses, who had invested much faith in a particular philosophy.

LYSENKO TAKES CONTROL OF GENETICS

During 1938 and 1939 the Soviet government's actions to control the research of the Academy of Sciences helped Lysenko to gradually strengthen his administrative power not only over agricultural science but also in genetics and biology generally.[25]

Lysenko was present when a meeting of the Soviet council of ministers decided in May 1938 that the research plan of the Academy of Sciences for the year 1938 could not be accepted, and a thorough reorganization of its work was demanded. One result of this decision was that a commission headed by Boris Keller should evaluate the work of Vavilov's genetics institute. The result was severe criticism of the institute's work and Vavilov's management of it. To compensate for the claimed weaknesses, Lysenko was invited to organize a new department within the institute.

Another important step was Lysenko's election to membership in the Academy of Sciences in January 1939. Among the candidates in biology were four VASKhNIL academicians, Kol'tsov, M. Zavadovskii, Tsitsin, and Lysenko. Strong political pressure was used to prevent the election of Zavadovskii and Kol'tsov. Kol'tsov was especially singled out as politically obnoxious. A report from the party organization at his institute said that Kol'tsov was not willing to recant his eugenic mistakes, that he protected people hostile to the Soviet regime, that he had foreign friends, and so on. A few days before the election there was also a letter in *Pravda* branding Kol'tsov as "fascist." The author is reported to have been Prezent (Babkov 1992, 451), and the letter was signed by a number of people working at the Academy of Sciences. The result was that Lysenko and Tsitsin were elected, not Kol'tsov or Zavadovskii, and consequently the position of classical genetics in the Academy of Sciences was seriously weakened.

At the same time the transfer of Kol'tsov's institute from the Ministry of Public Health to the Academy of Sciences was being considered as part of the ongoing reorganization of research under the academy. After the letter in *Pravda*, Kol'tsov, in a last desperate attempt to save his institute from Lysenkoist destruction, wrote to Stalin for help. He explained how Muralov and Iakovlev (both executed in 1937) had misinterpreted his earlier eugenics writings, how he had protested against sterilization, and how he had immediately broken off his interest in eugenics when the first signs of fascism emerged in Germany. He also reminded Stalin of how Gorkii had intervened to make him protect the institute when it was similarly threatened in 1932.[26] But his arguments did not help. Kol'tsov had to accept the transfer to the Academy of Sciences.

In March 1939 the Academy of Sciences set up a commission to review the work of the Institute for Experimental Biology. The chairman was A. N. Bakh, and Lysenko was one of the members. Bakh was a biochemist and one of the signatories of the abusive January letter in *Pravda*. He had supported Lysenko on a number of occasions. One session of the commission, on April 15, 1937, turned into a regular interrogation of Kol'tsov. Prezent played an active role, though he was not an official member of the commission. In spite of tough pressure, Kol'tsov was still unwilling to recant any eugenic mistakes, to "disarm himself" (*razoruzhit'sia*) (Babkov 1992, 452–53).

In April 1939 the presidium of the academy decided to dismiss

Kol'tsov from the directorship and to reform the institute. Part of the motivation was that its work was based on "metaphysical and idealist perversions of the theory of heredity and an antiscientific revision of Darwinism." The institute was retained as an institute of cytology, histology, and embryology, but the research program was changed. It was especially genetics that suffered (Gaissinovich and Rossianov 1989, 102).

Kol'tsov's ability to adapt his research organization to different ideologies and regimes has been emphasized by Mark Adams (1980). Kol'tsov understood how important it is to retain a set of researchers that are motivated to carry out a certain research program. Ideological principles and political declarations are not necessarily decisive as long as this social structure is preserved. For instance, when a number of the geneticists were arrested and exiled in 1932, Kol'tsov was able to recruit a new team of geneticists with the *vydvizhenets* (pushed ups) and party member Dubinin as leader. Dubinin, in contrast to Kol'tsov, emphasized ideological reasons in science policy. He was also among those who demanded Kol'tsov's public repentance on eugenics in 1939. This helped Dubinin to stay on as leader of an active research group in classical genetics until 1948. In spite of the victory in 1939, Lysenko was not able for another ten years to completely suppress classical genetics in the institute that Kol'tsov had created.

By the end of 1939 the directorships of the Institute for Genetics of the Academy of Sciences and of VIR were the only leading positions in Soviet genetics that Lysenko did not control. Both these positions still belonged to Vavilov. But on August 6, 1940, he was arrested and charged with espionage or collaboration with foreign powers. Later that autumn some of the key anti-Lysenkoist scientists of VIR were also arrested. Kol'tsov died of heart trouble in December 1940.[27] Iogann Eikhfel'd, the Lysenkoist plant cultivator from the Arctic, became the new head of VIR. Lysenko himself took over the directorship of the academy's Institute of Genetics and fired most of its scientific staff. In early 1943 Vavilov died from malnutrition and mistreatment in a Saratov prison.

EPILOGUE: FROM TEMPORARY OPTIMISM TO FINAL ROUT OF GENETICS

After the series of defeats in 1938–1940, the pressure on genetics eased temporarily during and immediately after World War II. Research in genetics continued at the academy's Institute of Experimental Biology under Dubinin and in a number of university departments. Zhebrak continued his work as head of genetics at the Timiriazev academy, the central institution for higher agricultural education.

As the war ended in 1945, geneticists took an optimistic view of future developments and attempted to push back Lysenko's influence and power. His position was not only being undercut by developments in agricultural and biological science. Lysenko's brother had defected to the enemy during the German occupation of the Ukraine. Nikolai Vavilov's brother, the physicist Sergei Vavilov, became president of the Academy of Sciences after Komarov died in December 1945. The common struggle against Germany and Japan had revived contacts with England and the United States.

Zhebrak led the campaign to restore Soviet genetics. Around the beginning of 1945 he wrote to Georgii Malenkov arguing the need to restore the activity and international status of Soviet genetics from the destruction that Lysenko had caused. Malenkov was at this time one of the top bosses in the secretariat of the party Central Committee. Zhebrak enclosed two articles about Soviet genetics published in the American journal *Science*, one by Leslie Dunn and the other by Karl Sax, professor of cell biology at Harvard University (Dunn 1944; Sax 1944). Both praised the achievements of Soviet genetics. It had "outstripped the Germans in this field even before the advent of Hitler," wrote Dunn.

But Sax protested against Dunn's rosy picture of the situation for Soviet genetics. He had not mentioned Lysenko and the troubles of Soviet genetics since the middle of the 1930s.[28] "Lysenko and his associates seem to have convinced the political authorities," wrote Sax. Since 1939 Soviet journals "have been filled with articles by Lysenko's disciples, but we hear nothing from Vavilov, Karpechenko,

Navashin, and the many other able scientists who are responsible for building the foundations of Russia's plant-breeding program." According to Sax, "the subservience of science to social and political philosophy" was a serious problem that the Western public should know about—a problem that was not confined to Russia. "It could happen here," he warned.

Zhebrak's communication to Malenkov was well received. From September 1945 to April 1946 Zhebrak worked as head of a section in the Central Committee's directorate of propaganda and agitation.[29]

Zhebrak replied to Sax in *Science*, claiming that his fears about suppression of science in the Soviet Union were quite unfounded. Zhebrak's article had been written on request of the "Antifascist committee of Soviet scientists" and had been passed on through the government information service. This origin set its mark on the article. "The Soviet government has never interfered in the discussions of genetic questions which have now been raging for some ten years," declared Zhebrak. He also claimed that Lysenko's directorship of the Institute of Genetics "does not mean that other schools of Soviet geneticists are in any way hampered in their work." Sax had not understood that dialectical materialism "is based on real facts and never denies them," and that therefore "science can be free in a centralized socialist state, which Dr. Sax wrongly calls totalitarian." Not only did Zhebrak's article paint a much too rosy picture of the situation for Soviet genetics, but it also contributed to upholding the myth of Lysenko's practical success by claiming that as "an agronomist, academician Lysenko has put forward a number of practical suggestions which have been of great value to the Soviet government."

Though Zhebrak was a loyal party member, he hardly meant every sentence he wrote, and he certainly did not write all he meant that was essential to a complete picture of the situation. Nevertheless, within the bounds set by current demands for loyalty to his country and its political system, Zhebrak did make some important points against Lysenkoism. For instance, he claimed that Mitin's speech at the 1939 conference expressed only Mitin's own views "and not in any way the viewpoint of the Soviet government." Coming from a person sitting in the Central Committee's division of propaganda and agitation, this was a warning to Lysenkoists that their views were being seriously challenged within the government. And in conclusion, Zhebrak praised international collaboration in science. He hoped that misunderstandings about "the basic ideas of our country" and the development of its science "will be speedily dispelled, and that in the future the scientists of the two countries will progress together in an atmosphere of mutual understanding" (Zhebrak 1945).

However, communication with geneticists in the West at this time was by no means free and unrestrained. The International Genetics Committee made unsuccessful efforts to place the Eighth International Congress of Genetics in the Soviet Union. But "communications sent to various institutions in Russia received no answer after a considerable period of waiting," and it was decided to locate the congress in Sweden.[30] It was held July 7–14, 1948, with no genuine geneticists, but only a couple of Lysenkoists, participating from the Soviet Union.

Lysenko and his supporters were distressed by the increasing activity of the geneticists. In March 1947 Zhebrak was named president of the Belorussian Academy of Sciences. During the same month there was also a genetics conference at Moscow University. The Lysenkoists were unable to stop this conference or to make serious reprisals against the organizers. On the contrary, a report "on the situation in VASKhNIL," unfavorable to the interests of Lysenko, was prepared by a commission of the Central Committee secretariat. The Academy of Agricultural Science was not working well, said the report. For instance, Vice President Tsitsin had practically quit his job because of disagreements with Lysenko. The commission recommended the election of a number of new academicians in VASKhNIL as part of a program to revitalize the organization.[31]

At the end of April, Zhebrak and S. I. Alikhanian, a geneticist at Moscow University, wrote to Andrei Zhdanov, the party chief of culture and science, drawing an analogy between genetics and atomic physics, that is, between research on the gene and the atom. If there was disagreement in physics concerning the possibility of liberating nuclear energy, "with one school simply rejecting any possibility of progress in the analysis of atoms and constructing absurd and fantastic projects, and the other, based on precise scientific facts, demanding development of the theory of the atom, then, obviously, the victory of one or the other point of view would determine the success of the work." Having in mind the atomic bomb, which the Soviet Union was just then successfully developing to catch up with the West, the thrust of the analogy was clear: future success in genetics demanded a theory of gene structure and function, but the Lysenkoists more or less denied the very existence of the gene. Zhebrak and Alikhanian asked Zhdanov for the difficulties of Soviet genetics to be considered by the Central Committee. They wanted a normal development of genetics to be ensured by eliminating "the irregularities created by the activity of academician Lysenko."[32]

But the start of the cold war, with campaigns against internationalism and subservience to the Western capitalist countries and their culture, gave the Lysenkoists a platform for counterattack. Zhdanov had the central responsibility for these campaigns. On September 2, 1947, *Pravda* published an article in which Zhebrak and Dubinin were accused of writing antipatriotic articles[33] in a foreign journal and not defending Lysenko from the attacks he had been subjected to in the foreign press. This was disloyal behavior for a Soviet citizen. The two ought to be judged by a "court of honor."[34]

The geneticists and their supporters reacted strongly. But they did not have access to the press and were reduced to writing letters to influential political leaders. There was no public forum for this debate. Arguments were minimally confronted with knowledgeable criticism. Wishful thinking based on superficial and one-sided information had free rein outside the scientific community.

The plant physiologist D. A. Sabinin wrote to Zhdanov that solidarity with Lysenko's genetic writings was not possible anymore. His book *Heredity and Variation* had now been translated into English and was well known in the West. A Soviet biologist reading Lysenko's statements about heredity as a power of the cell as a whole, that "every drop of protoplasm possesses heredity," and so on, could only close the book with indignation and shame. If Zhdanov supported Lysenko, it would simply prove the false accusation from Sax, that science was not free in the Soviet Union. The independent and critical attitude toward Lysenko was precisely what made Zhebrak's reply to Sax so convincing and valuable, argued Sabinin.[35] Zhebrak himself wrote a letter to Zhdanov with a plea to meet him personally to have the matter cleared up and to help in publishing an explanation in *Pravda*.[36] Lisitsyn in another letter to Zhdanov reacted violently against the "coarse demagogical tone" of the articles accusing Zhebrak of antipatriotism. "It is time that such licentious authors are called to order."[37]

The outspoken criticism in these letters did not help much. Lysenko strengthened his position in the autumn of 1947 as some of his most active critics were politically compromised. Zhebrak was sentenced with a public reproof (*obshchestvennyi vygovor*) in November and removed from the presidency of the Belorussian Academy of Sciences. Dubinin narrowly escaped a court of honor with the help of top officials in the Academy of Sciences.[38]

The conflict over genetics escalated further in early 1948. At a conference on the problems of Darwinism organized at Moscow University in February, Lysenko's views were again subjected to sharp criticism and received only feeble support. There was not the same youthful revolutionary atmosphere at the universities as in the 1930s. Time and publicity were not working in favor of Lysenko in the same way they had ten years earlier. The majority in the biological section of the Academy of Sciences, led by the evolutionist I. I. Schmalgausen, was systematically working to push Lysenko out of his positions.[39]

At the beginning of 1948 Iurii A. Zhdanov became head of the section of science in the directorate of propaganda and agitation (*Agitprop*) of the Central Committee. As the son of Andrei Zhdanov he was personally acquainted with Stalin. According to Iurii Zhdanov's own account, published twenty-five years afterward, it was Stalin personally who recruited him to this work, somewhat against the advice of his father. Iurii Zhdanov reports the content of two conversation in which Stalin presented him with essential knowledge about Soviet science and its politics, biology in particular. Stalin's own achievements in gardening were an important part of the conversation, according to the account of the younger Zhdanov. For instance, how Stalin had successfully introduced useful plants to new environments was in conflict with expert opinions. He also explained to young Zhdanov how classical geneticists had overlooked the inheritance of acquired character and had become stuck with their theory of an unchanging genetic matter. Lysenko was an *empirik* and weak on theory. A strengthening of theory in Russian science was now necessary. Stalin was worried about his country's lacking success in making new discoveries. There was no less inquisitiveness in Russian science than abroad, but better organization was needed (Zhdanov 1993, 69–71).

In his new job Iu. Zhdanov was faced from the start with the controversy in biology and conducted discussions with many scientists, including Dubinin, Tsitsin, and Zhebrak. Soon he complained in a letter to Stalin himself that Lysenko as president of VASKhNIL had disrupted the development of a new variety of dandelion for the production of rubber.[40] Considering the importance of the production of natural rubber in our country, one should "give serious attention to the work of Prof. Navashin and help him," concluded his letter.[41]

The young Zhdanov launched a more general criticism of Lysenko in a public lecture on April 10, 1948, entitled "Controversial Questions of Contemporary Darwinism." Iu. Zhdanov's argument struck what can be called a middle course, but it was in fact the only kind of opposition now possible. He accepted in full the officially established science policy doctrines, including the practice criterion, but tried to use it against Lysenko.

The younger Zhdanov explained in his lecture[42] that the debate between the two schools of genetics was taking place within the framework of Soviet biology, and neither of them could be called "bourgeois." He criticized Lysenko on several points, for not taking the methods for multiplying chromosome numbers seriously, for rejecting the use of hybrid corn, for his unfulfilled promises of creating new useful plant varieties in two to three years, and so on. But he also repeated the usual praise of Lysenko and the standard methodological criticism of the geneticists. Lysenko had made great contributions to practical agriculture, for example, vernalization.

The geneticists were obsessed with fruit flies, believed in unchanging genes, and suffered from a divorce of theory from practical problems. It was Lysenko's attempts to suppress other schools and obtain a monopoly that Zhdanov criticized, not the general fruitfulness and legitimacy of his approach. He was critical of the way Mitin and other philosophers had intervened in favor of Lysenko, but he upheld their theoretical principles.⁴³

Lysenko listened to the lecture from Mitin's office, situated in the same building where Iu. Zhdanov spoke. He complained to Stalin and A. Zhdanov that the lecture had repeated the slanderous claims of the anti-Michurinists. Coming from a high official in the Central Committee, people were likely to believe this slander. His position as president of VASKhNIL had thus become very difficult, wrote Lysenko. He also complained that the speaker had not once called on him to discuss the matter before he brought it before the public. A month later, in early May, Lysenko took the calculated risk of asking the minister of agriculture to be relieved from the post as president of VASKhNIL.⁴⁴

Apparently, negative attitudes toward Lysenko were strong in the party Central Committee. The younger Zhdanov hardly acted without the knowledge and approval of his father. And his immediate superior, the head of *Agitprop*, Dmitri Shepilov, was certainly well informed. But Stalin was not motivated to drop Lysenko. His sympathy for Lamarckian views was well known. Iurii Zhdanov in his anti-Lysenko lecture mentioned how Stalin himself had stressed the positive aspects of the inheritance of acquired characters in his work "Anarchism or Socialism" (Zhdanov 1993, 77). Stalin the romantic gardner is also revealed in the story about branched wheat.

Wheat varieties with many spikes on each straw, called branched wheat, had persistently fired the imagination of peasants. Such a plant immediately suggested a manifold yield. But plant breeders had in vain tried to combine branching with the ordinary important qualities of cultivated wheat to create high-yielding varieties. In 1946 the Soviet Union suffered a bad drought, and on New Year's Eve Stalin called Lysenko to the Kremlin to take on the task of creating a new and miraculous branched wheat.⁴⁵ Results from the first year of experimenting are described in a personal research report to Stalin in October 1947. The tone is optimistic, with great hopes for the future, but the results were in no way sensational. Stalin's reply complained a little that the wheat had not been given optimal conditions satisfying the need for heat as well as moisture of such a southern variety. But in general he encouraged Lysenko's research program, especially his ideas of hybridizing the branched with other varieties. The last paragraph of Stalin's letter stated in clear terms that Michurinism is the only valid scientific basis for biology: "Weismannists and their followers, who reject the inheritance of acquired characters, no longer deserve to be tolerated. The future belongs to Michurin." Lysenko's report and Stalin's reply were sent to all members of the Politburo with his personal note that the problems stated in the report would be discussed in the Politburo in due time (Vavilov 1998). For a while branched wheat gave rise to sensational newspaper stories, and Lysenkoists continued for some years to claim new high-yielding varieties (Soyfer 1994, 179).

At a meeting of the Politburo in early June, Stalin himself launched a counterattack chastising Iurii Zhdanov and his immediate superior, Dmitri Shepilov, for their critical attitude toward Lysenko. Iurii Zhdanov wrote a letter of repentance to Stalin, later published in *Pravda*. In early July, Andrei Zhdanov had to cooperate with Mitin in drafting a resolution for the Central Committee, "On the Situation in Soviet Biological Science." Here "progressive, materialist, Michurinist biology" was opposed to "reactionary-ideological, Mendel-Morganist" biology. As adherents of the latter were named Shmalgausen, Zhebrak, Dubinin, Kol'tsov, Serebrovskii, and Navashin. "In science as in politics, contradictions are solved not through reconciliation, but through open struggle," the declaration asserted. Iu. Zhdanov's attempt at eclectic reconciliation was a mistake. The "followers of Mendelism-Morganism have more than once been warned that their direction in biology is foreign to Soviet science and leads into a blind alley," the declaration continued, but they "continue resisting and expanding on their mistaken views. . . . Such a situation can no more be tolerated."[46]

As late as July 16, Konstantinov wrote a long letter to Stalin repeating the main points in the criticism of Lysenko's agricultural science and techniques.⁴⁷ But the matter had already been decided. Instead of the Central Committee adopting a resolution, however, a new VASKhNIL congress called "The Situation in Biological Science" was organized, with Lysenko giving the introductory address. A number of new academicians of Lysenkoist opinion were appointed shortly before the congress, and there was only weak opposition to Lysenko's views. Zhebrak defended his polyploid grains. Alikhanian tried to disentangle neo-Darwinism from Lysenkoist misinterpretations. P. M. Zhukovskii likewise attempted to explain his position on chromosomes and the influence of the environment. Lisitsyn had died some months earlier, and Konstantinov sat silently in the audience. Boris Zavadovskii held on to his "third position" from 1939. But he now concentrated on the disagreements with Lysenko. Both Lysenko and Mitin were simplifying the situation in an illicit way. For instance, they attributed to him some of the same idealist mistakes as the Mendelians. Their extreme one-sidedness did not give room for an open scholarly debate, B. Zavadovskii now complained.⁴⁸

The discussion as a whole was dominated by abstract and somewhat outdated biological theorizing and simplistic philosophical arguments, interspersed with references to Engels, Marx, Lenin, Stalin, Darwin, Timiriazev, and the like. No philosophers participated in the debate, but the conceptual framework and the science policy principles of the debate were in accordance with the official Soviet philosophy of science.

This Lysenkoist argument was essentially what Mitin and the other philosophers had claimed at the 1939 conference pursued to its logical conclusion, assuming that Michurinism was in fact of superior practical usefulness. However, as Krementsov has underlined, this disastrous turn in the fortune of Soviet genetics happened just as the cold war was escalating to its peak. The Berlin Blockade started on June 24, 1948 (Krementsov 1997, 158–59). Without this background of high international political tension, it is likely measures against classical genetics would have been much less draconian.

Only toward the end of the congress did Lysenko tell the audience that his report had already been examined and approved by the

Central Committee. This led to immediate recantations by some of Lysenko's main opponents and became the signal to close down research and teaching of classical genetics all over the Soviet Union.

It is often assumed that the principle of *partiinost* and the opposition of proletarian to bourgeois science was the ideological basis for Lysenkoism. But historians have revealed how Stalin personally corrected and edited Lysenko's talk on this point. He systematically deleted all references to "bourgeois biology." Lysenko's claim that "any science is class science" was likewise rejected. In the margin of the manuscript, Stalin wrote: "Ha-ha-ha!!! And mathematics? And Darwinism?"[49] He apparently saw how radical social relativism was not a tenable position. It could be useful as a subversive political instrument, but for someone in a governing position, the primary problem is to decide which theory is right and which is wrong. The differentiation between progressive and practical Soviet science on the one hand, and metaphysical and sterile Western science on the other, was more adequate in indicating clearly why the one was correct and the other mistaken. The correct theory, or at least the most correct one, was needed as a sound basis for positive and constructive actions of a responsible government.

Stalin's critical attitude toward radical social constructivist views of scientific knowledge were confirmed by his intervention into the controversy in linguistics a few years later. In the words of Joravsky (1970, 151), Stalin "not only removed politically imposed crackpots from control of the community of linguists; he restored genuine self-government to that community." So perhaps we could say that even in the midst of the most notorious example of the political suppression of scientific autonomy, the key person was well aware that such autonomy is necessary for a well-functioning modern political regime. The problem was not so much the lack of theoretical insight as of stable social institutions that incarnated the theory and could act on it when necessary.[50]

Iurii Zhdanov's role in precipitating the 1948 showdown between Lysenkoists and geneticists apparently did not harm his sympathy with Stalin. He stayed on in the top party hierarchy and even got married to Svetlana, the daughter of the great leader. As it turned out, the provocation may have been quite convenient for Stalin. Iurii Zhdanov's philosophical principles for science policy were quite similar to Stalin's own, even if the young and inexperienced Zhdanov had drawn a wrong conclusion. The influence of the principles continued. Iurii Zhdanov in the 1990s emerged as a strong candidate for the post as director of the academy's Institute for the History of Natural Science and Technology.

Not only did the Soviet government under Stalin's leadership stage the 1948 VASKhNIL session, its belief in the truth of Michurinism was also demonstrated by making the congress an international showcase. A verbatim report of the proceedings was translated into Western languages and published in large editions. The character of the Lysenkoist arguments as deeply contrary to the traditions of science was thus effectively exposed for the whole world. The effect on Western intellectual sympathizers was deep and shocking. It was a most amazing example of a government shooting itself in the foot.

Thus Lysenko in 1948 won a complete victory in his struggle with the geneticists. But this stimulated rather than broke the resistance of the Soviet scientific community as a whole. In other central fields of natural science, such as physics and chemistry, similar attempts to establish a theory in accordance with the official doctrine of dialectical materialism were warded off. Physicists, chemists, and mathematicians supported their biologist colleagues and helped build the base for a counterattack in due time.[51]

Already before Stalin's death in 1953 articles critical of Lysenko's views on natural selection started to appear, especially in the main botanical journal, *Botanicheskii Zhurnal*. But the hope that after Stalin's death Lysenko and his supporters would quickly be deposed from their positions of power did not materialize. They were still favored by the political leadership. Nikita Khrushchev had a high estimation of Lysenko's agricultural expertise. This lasting support demonstrates that the preference for Lysenko in 1948 was not merely a whim of Stalin but based on a widespread sympathy and belief in his ideas. Only when Khrushchev was removed from office in 1964 were the Lysenkoists ousted. Lysenko was not stripped of his membership in the Academy of Sciences, however, and he kept his own research institute on the Lenin Heights in Moscow until he died in 1976.

NOTES

1. TsGAE, f. 8390 o. 1, ed.khr. 1136.
2. See, for example, the exchange between Vavilov and Lysenko at the genetics conference, October 7–14, 1939, *SGS*, vol. 2, pp. 413–21.
3. See J. W. Pincus, "The Genetic Front in the U.S.S.R.," *Journal of Heredity* 31 (1940): 165–68. The article is largely based on a report in *Iarovizatsiia* 2, no. 23 (1939): 117–22.
4. For political and administrative circumstances around the conference see Krementsov (1997, 71–76).
5. A general report of the conference and several of the lectures are printed in *Pod Znamenem Marksizma* 10 and 11 (1939). An English translation of some of the talks, partly in abbreviated form, was published in *Science and Society, a Marxian Quarterly* (New York) 4 (1940): 183–233. There also exists a stenographic report of the conference (*SGS*).
6. *SGS*, vol.1, pp. 1–9.
7. For instance, when Mendel crossed his peas, yellow seed and green seed gave plants with only yellow seed in the first generation of offspring. Yellow was "dominant" over green. Crossing within this first generation would then give the characteristic 3:1 distribution of offspring in the next generation. For every three plants with yellow seed, there would on an average be one with green. This ratio was interpreted as the result of hereditary factors embedded in each cell of the organism.
8. *SGS*, vol. 1, p. 90.
9. *SGS*, vol. 2, pp. 67–68.
10. *SGS*, vol. 2, p. 75.
11. *SGS*, vol. 2, pp. 56–57.
12. This was part of Lysenko's theory about hybridization through grafting. The base on which a scion is grafted can influence the heredity of seed grown on the scion, thus acting as a mentor.
13. *SGS*, vol. 1, p. 103.
14. *SGS*, vol. 1, p. 102.
15. *SGS*, vol. 1, p. 206.
16. *SGS*, vol. 1, pp. 221–22. The names that Zhebrak mentioned are omitted from the stenographic report.

17. *SGS*, vol. 2, pp. 199–200.
18. *SGS*, vol. 2, p. 206.
19. *SGS*, vol.1, pp.194–95.
20. *SGS*, vol. 1, p. 33.
21. "Ot redaktsii" (from the editorial board), *PZM* 10 (1939): 146.
22. *PZM* 11 (1939): 121. L. I. Govorov was probably referring to Konstantinov and Lisitsyn.
23. This was the Russian translation of the internationally renowned *Principles of Genetics* by E. W. Sinnot and L. C. Dunn. At the time this book was being widely used in Soviet universities and other institutions of higher education.
24. *SGS*, vol. 2, pp.76–80. Johannsen (1857–1927) was at first very critical of eugenics, but in his later years, especially after 1920, he became an active promoter of a "reform" type of eugenics.
25. For descriptions of this development see Soyfer (1989, 299ff; 1994, 121ff) and Krementsov (1997, 61ff).
26. The complete text of the letter is published in Gaissinovich and Rossianov (1989).
27. For details concerning Vavilov's arrest and connected events see Soyfer (1989, 323–30).
28. Dunn's article was the text of a talk he had given in "the Science Panel of the Congress Celebrating the Tenth Anniversary of American-Soviet Friendship," New York, November 7, 1943." See Dunn (1944, 65fn).
29. Upravlenie propaganda i agitatsiia TsK VKP(b). See *ITsKKPSS* 4 (1991): 125–30.
30. *Proceedings of the Eight International Congress of Genetics, 7th–14th of July, 1948, Stockholm*, ed. Gert Bonnier and Robert Larssoneds, supplementary volume of *Hereditas*, 1949.
31. See materials published in *ITsKKPSS* 4 (1991): 125–32.
32. Letter from A. Zhebrak and S. Alikhanian to A. Zhdanov, April 28, 1947, *ITsKKPSS* 6 (1991): 157–58.
33. The reference was to Zhebrak's 1945 reply to Sax, which I have discussed earlier, and to an article by Dubinin published in early 1947. In this article Dubinin described work in "theoretical genetics" by Soviet scientists. He mentioned several people who had been purged, among them Vavilov and Karpechenko. Dubinin even referred to N. V. Timofeev-Ressovsky, who had worked under the Nazi government in Berlin throughout the war and was then imprisoned by the Soviets. Lysenko on the other hand was not mentioned at all.

34. The article was built on material from articles in Lysenko's journal *Agrobiologiia*, the former *Iarovizatsiia* 5-6 (1946). Similar articles were also published by *Literaturnaia Gazeta*, where Mitin had just become editor for science and technology, and by *Sotsialisticheskoe Zemledelie*.

35. D. A. Sabinin to A. Zhdanov, September 8, 1947, *ITsKKPSS* 6 (1991): 164-65.

36. A. Zhebrak to A. Zhdanov, September 5, 1947, *ITsKKPSS* 6 (1991): 167-68.

37. P. I. Lisitsyn to A. Zhdanov, September 10, 1947, *ITsKKPSS* 6 (1991): 166.

38. *ITsKKPSS* 6 (1991): 172.

39. See, for instance, Rossianov (1991, 15-16) and the materials and commentaries published by V. Esakov, S. Ivanova, and E. Levina, "Iz instorii bor'by s lysenkovshchinoi" (from the history of the struggle against Lysenkoism), *ITsKKPSS* 7 (1991): 109-21.

40. The creation of a dandelion suitable for rubber production was a first priority project during the world war but had of course lost some of its urgency when the war ended. The promising variety that Lysenko interfered with was a tetraploid, that is, it had a double set of chromosomes, and was thus developed on the basis of chromosome theory and cytological methods that Lysenko had been very critical of, for instance at the 1939 genetics conference. Much of the work on dandelion for rubber production had been carried out in Belorussia where Zhebrak was for a while president of the Academy of Sciences.

41. Letter from Iu. A. Zhdanov to I. V. Stalin, February 24, 1948, *ITsKKPSS* 7 (1991): 110.

42. Published in 1993 in an abbreviated version edited by Iu. Zhdanov himself. Even at this time he did not distance himself from the principles of Soviet science policy of the 1940s. (Zhdanov 1993.)

43. For the content of Iu. Zhdanov's lecture, see also Soyfer (1989, 386-89), Rossianov (1991, 16-19), and E. Esakov, S. Ivanova, and E. Levina in *ITsKKPSS* 7 (1991): 109-13.

44. Letter from Lysenko to Stalin and A. Zhdanov, April 17, 1948; letter from Lysenko to the minister of agriculture, I. A. Benediktov, May 11, 1948. Both letters are printed in Soyfer (1989, 390-91).

45. For more details about this story see Soyfer (1994, 177-79).

46. See *ITsKKPSS* 7 (1991): 112-13. The declaration was first drafted by D. T. Shepilov and Mitin, corrected by A. Zhdanov, who included the criti-

cism of his own son among other things, and then sent by Zhdanov and Malenkov to Stalin as well as to the other members of the Politburo.

47. Letter from P. N. Konstantinov to I. V. Stalin, July 16, 1948, *ITsKKPSS* 7 (1991): 113-19.

48. *The Situation in Biological Science* (1949), pp. 339-60.

49. See Rossianov (1993, 732).

50. Or, to keep to a terminology of methodological individualism: institutions that make it possible for individuals to act effectively so as to gain acceptance for rational arguments.

51. For a general account of the developments within the Academy of Sciences see Vucinich (1984). A description of the struggle from a biologist's point of view has been given by V. Aleksandrov (1987-1990) in a series of articles on "The Difficult Years of Soviet Biology." Another important participant and chronicler was A. A. Liubishev (1991).

Chapter 10.

Concluding Discussion: Why Did It Happen?

How did the promotion of poor science and the subsequent suppression of good science come about, leading up to the scandal of 1948? What were the mechanisms and driving forces in this process?

Lysenkoism grew in social milieu under the influence of political tyranny and mass terror. But an explanation that refers primarily to the political suppression of genuine science is not enough. Such accounts set Soviet Lysenkoism too much apart from normal science and prevents interesting comparison with contemporary as well as present science in the West. I have shown how it was enthusiasm and dedication that marked Lysenko's initial research, not complete incompetence and an unscientific attitude, "pseudoscience" or "lying science" (*lzhenauka*), as it has often been called. An important contribution to such relative normalization of Soviet science is the new institutional accounts (Krementsov 1997; Kojevnikov, 1996, 1998, 2004) describing the social systems that set the framework for individual careers, entrepreneurship, and negotiations on science policy. Such extremely rapid growth as the Soviet research system experienced under Stalin will in any circumstances produce radical change. Admittedly, the social and political circumstances were special, but Stalinist science nevertheless remains a grandiose and unique social experiment of general interest to science policy.

I have focused on the content of a particular sector of Stalinist science, the intersection of plant physiology, breeding, and genetics. My account has shown how the formulation of scientific problems and research programs developed, gradually letting the poor science drive out the good. A science policy strongly promoted by the government, but also broadly accepted by the scientific community and enthusiastically carried out by leading scientists of high international reputation, prepared the ground for Lysenkoism. The suppression of genetics was the most notorious, an extreme example of a typical phenomenon in the politics of Stalinist science. In the social sciences and the humanities the ideological standardization came much more quickly and radically. In the hard natural sciences such as chemistry and physics, it never got very far. But biology became the striking example of a science in which traditional criteria of evidence and argument, objectivity and truth, were overruled by normative political considerations. I have concentrated on the first phase of the drama, the *rise* of Lysenkoism. It is this phase that most readily answers the question: How was it possible that a primitive and retarded biological science came to dominate the Soviet Union for decades?

AUTONOMY OF SCIENCE

When the American sociologist Robert K. Merton developed his "ethos of science" in the late 1930s and the 1940s, it was primarily under the impression of events in Nazi Germany. Soviet Russia was a major ally in the war against Germany and only emerged fully as the second great oppressor of scientific freedom after the end of World War II. Merton's norms for the institution of science—universalism, communism, disinterestedness, and organized skepticism—described a liberal ideal of science. *Universalism* referred to the universal validity (truth) of the knowledge produced; *communism*, to the free access, that the knowledge was public property; *disinterestedness*, to the detachment from specific practical uses of the results, that is, particular interests should not affect the conclusions; and *organized skepticism* referred to the tradition of systematic critical scrutiny, no result is accepted before all objections have been carefully considered (Merton 1942).

Merton's norms describe an ideal for the acquisition and management of general theoretical knowledge, what is traditionally called *basic science* as opposed to *applied science*. The latter is also often called *technology*. At the core of this Mertonian ideal lies trust in the possibility of objective knowledge valid for all cultures and all times. Not absolute objectivity or absolute universality, but substantial enough for sound development of technological practice as well as human culture. In Merton's view there was a strong interdependence between such science and liberal democracy. But he also warned that this science was dependent on cultural conditions that gave it sufficient prestige to motivate and fund capable persons for scientific careers: "changes in institutional structure may curtail, modify, or possibly prevent the pursuit of science" (Merton 1938, 254).

Such a liberal and internationalist ideal of a science, combining technological fruitfulness with beneficial cultural development, was what people like Nikolai Vavilov had been socialized into in the years before the October revolution. But it was not trusted by Marxists. To them it expressed an ideology suited to perpetuate bourgeois hegemony. In Marx's view such a contemplative ideal of science was not helpful in transforming society. His eleventh Feuerbach thesis was a favorite quotation under the rule of Soviet communism: "Die Philosophen haben die Welt nur verschieden interpretiert; es kommt darauf an, sie zu verändern" (Philsophers have only interpreted the world differently, what matters is to change it). This activist and voluntarist attitude toward scientific knowledge reached a high point in the Stalinist doctrine of the unity of theory and practice, expressed in the practice criterion of truth and the *partiinost* of science. This doctrine acted as a strong bid from the patron to take over governance and eliminate scientific autonomy.

One way of accommodating the dilemma of integrity versus usefulness is to introduce a distinction between basic (theoretical) and applied (practical) science. This distinction is expressed in varying terms, for instance, as science versus technology. The underlying idea is a fundamental difference between knowledge and action.

Human rational action presupposes both a goal, something of value to be reached, and sufficient knowledge about the situation. The same knowledge in the same factual situation can lead to quite different actions if values are different. While basic or pure science has knowledge of general validity and common interest as its goal, applied science or technology is oriented toward practical tasks, from curing disease and preventing hunger to enhancing entertainment or the winning of wars, or committing crimes for that matter. All the latter tasks depend on reliable general knowledge to succeed, but are not primarily aimed at extending it.

The distinction does not provide a sharp divide between two separate categories of science or research. Much research as well as other scientific activity is hard to categorize as either basic or applied and is best seen as some of both. But like the vague distinction between night and day is essential for organizing our lives, the vague distinction between basic and applied science (or something equivalent) is essential to organizing science. Or so argue its proponents. The implication is that there is one area of science where the activity is governed in a (relatively) autonomous manner by scientists themselves, while other areas are under the governance of practical agencies of state or private industry and geared toward their needs for knowledge. The reason why the state as well as private enterprise should be willing to fund basic science is its long-term benefits to the whole of society. Two arguments, the technological and the cultural, have been much used. First, a science governed by scientists is most effective in making fundamental new discoveries and thus in supplying radical new technical possibilities. It takes intimate knowledge of the research front to see and judge what the revolutionary possibilities are. Second, a science aimed at knowledge that is valid for everybody is best suited to further human culture, increasing humanity's understanding of its own situation and thus improving its ability to make wise decisions. Especially such a relatively autonomous basic science with high authority in general questions of knowledge is the best way of securing sound methods in the various activities of applied science where political and commercial pressure is high. This is in brief the enlightenment rationale behind the science policy distinction between basic and applied science, or science and technology.

A sound defense of scientific autonomy must avoid two evils, according to Merton. On the one hand is the tendency to integrate science so closely with the rest of society that it loses the ability to perform its tasks of revealing truth and criticizing prejudice. On the other hand is the tendency to sever the bonds to such an extent that science loses sensitivity or relevance to ordinary human concerns. Such isolation can foster the hubris of scientism as well as the smug self-indulgence of "pure science." Merton in 1938 paid much attention to the danger created by taking the ideal of "pure science" too literally. On this point he was in line with left-wing critics of the academic "ivory tower." "Doubt about the blessings of technological progress is growing, and scientists have to take this problem seriously. They can no longer naively assume that the social effects of science *must* be a blessing in the long run." Such an assumption involved "the confusion of truth and social utility which is characteristically found in the nonlogical penumbra of science" (Merton 1938, 263).

Only when the 1942 paper was republished in 1949 did Merton, in an added footnote, explicitly refer to the Soviet Union by pointing out how the norm of universalism was rejected in the Soviet philosophy of science.[1] Lysenkoism was not mentioned, though it had surely been present in Merton's thoughts. There is no doubt that many scientists in the 1930s drew the parallel between Nazi and communist suppression of science and democracy. The American geneticist Bentley Glass has told that in Berlin in 1933 he tried in vain to make H. J. Muller see "that there was essentially no difference between the domination of science by the Nazi leaders and by the Soviet leaders. Hitler and Stalin were equally untrustworthy and viewed science only as a basis for technological improvements in military arms and economic resources" (Glass et al. 1990, 416). For Otto Mohr, as chairman of the international committee of genetics, the similarity of communist Russia and Nazi Germany was clear. He was fighting a war on two fronts for the freedom of science from political censorship and dictate. And by the spring of 1937 his judgment was clear: the suppression of scientific freedom in the Soviet Union had gone so far that it was no more responsible to hold an international conference of genetics there.

In spite of strong ideological standardization, the liberal ideal of science did surface during the 1930s Soviet policy debates over genetics. In 1936 Mikhail Zavadovskii argued provocatively that genetics *is* a theoretical discipline: Lysenko is quite right that "contemporary genetics is isolated from practical inquiry," but this isolation is unavoidable because genetics is a science about "objectively existing forms of change in matter" and not about "measures to reach set goals" (Zavadovskii 1936, 95).[2] The article was headed by an editorial comment that it was being published "for discussion" in connection with the prospective International Congress of Genetics, and it would be followed by articles from Lysenko, Shlykov, and others. One purpose of the editors was to exhibit the dubious ideological basis of Lysenko's critics.

WESTERN REACTIONS AND THE FEEDBACK TO SOVIET POLITICS OF SCIENCE

The autonomy of science was an internationally hot issue in the politics of science in the 1930s and 1940s. In Britain the rise of Lysenkoism was a main concern in debates between the group of young leftists inspired by the visit of Bukharin et al. in 1931 on one side and liberals such as the émigré Hungarian chemist Michael Polanyi and the Oxford zoologist John Baker on the other. In the euphoria after the 1931 visit, Bernal formulated a choice that would seem obvious to young idealists of left-wing persuasion: "Is it better to be intellectually free but socially totally ineffective, or to become a component part of a system where knowledge and action are joined for one common social purpose?"[3] But there were doubts about the Soviet situation also on the left. Haldane was concerned as eugenics was suppressed and human genetics came into difficulties at the beginning of the 1930s (Haldane 1932, 137).

In 1937 *Nature* commented briefly on the December 1936 conference. Modern genetic theory had been attacked for neglecting "the Marxian principle of the unity of theory and practice," and Lysenko had promoted Lamarckian methods for practical breeding.[4] The discussions were revealing of "the atmosphere in which scientific investigators in totalitarian countries have to live and work."[5] When the *London Times* in its annual survey of 1937 compared the fate of genetics in the Soviet Union with that of scientific anthropology in Nazi Germany, Joseph Needham assured that there was no reason for alarm. He had "firsthand evidence that work is going on in the genetical institutes in the USSR almost unaffected by the controversy." Lysenko's accusations against classical genetics for being "formal," should not be taken too seriously. The theories of both "the schools of Lysenko and Vavilov alike are subjected to the acid test of practice" (Needham 1938).

Both Needham and Haldane saw the difference between Lysenko's plant physiology and his genetics. Haldane characterized Lysenko as a "great Soviet scientist," referring to vernalization,[6] but he thought it most likely that Lysenko's general genetic theory was false.[7] However, some of the Lysenkoists' experimental results, for instance, on graft hybridization, might well be important scientific contributions. In any case science, including genetics, was so badly treated in the West that it was surely better off in the Soviet Union, according to Haldane. Right up to 1948 he insisted on seeing the Soviet genetics debate as a normal scientific controversy between two schools.[8]

In 1941 Baker organized the Society for Freedom in Science, with Polanyi as leading ideologist. In a lecture called "The Autonomy of Science" Polanyi in early 1943 depicted two "authoritarian" ideologies that posed a threat to free science, Nazism and communism. Both held that scientific teaching and research must be controlled by the state (Polanyi 1943, 30). State intervention in Soviet genetics and plant breeding had started around 1930 and was "definitely established" by the conference on planning of genetics and selection in June 1932. In the reports from the October 1939 conference he found "impressive evidence of the rapid and radical destruction of a branch of science" (Polanyi 1943, 33–34). Vavilov had appealed to the framework and standards of international science in his defense of genetics. But in Polanyi's view, Lysenko rightly answered by quoting Vavilov's own declaration from 1932 that the "divorce of genetics from practical selection" typical of the West must be eliminated in the Soviet Union. The methodological revolution that Vavilov supported in 1932 undermined the autonomy of science and therefore the legitimacy of the objections he raised in 1939 (Polanyi 1943, 34–35). Nevertheless, Polanyi's analysis ended on an optimistic note. Hopefully the Soviet government would realize that its plant breeding was "operating on lines which were abandoned as fallacious in the rest of the world about forty years earlier" (Polanyi 1943, 37).

In the Soviet Union this debate over the autonomy of science was forced underground. Only occasionally was autonomy defended in public by courageous and staunchly independent scientists like Michail Zavadovskii. However, the British debate demonstrates that the ideology of science that drove Soviet science politics was not dependent on terror and political suppression. It was a view that scientists of a socialist or left-wing persuasion shared internationally. A more specific and direct influence on internal Soviet developments came from the Imperial Bureaus of Plant breeding. We have seen how Mark Mitin at the genetics conference in 1939 referred to a review article by R. O. Whyte to prove the high international scientific standing of Lysenko (Mitin 1939, 151).

At the end of the Second World War, geneticists in the West were much concerned about the fate of their discipline and their colleagues in the Soviet Union. Information was scant; for instance, it took several years before it was clear that Vavilov was dead. To enlighten Western scientists and public opinion, Leslie Dunn and Theodosius Dobzhansky organized the publication in 1946 of an English version of representative work by Lysenko, *Heredity and Its Variability*. There were numerous reviews of this book and other worried articles in the Western scientific press, authored by Hermann Muller, Cyril Darlington, and others who had had close contacts to Soviet geneticists in the 1930s. But in the conciliatory atmosphere of the war and the immediate postwar period, criticism was muted.

Western scientists did not want to provoke political retaliation against their Soviet colleagues. Dunn thought the best strategy was to make Lysenko's ideas known in his own words, so that anybody could judge their value (Krementsov 1996; Harman 2003), while Dobzhansky wanted more open and direct criticism (Dobzhansky 1946, 5). It must also be remembered that the big weight on the conscience of the scientific community at this time was the atomic bomb. Peaceful reconciliation between the Soviet Union and the West was a prime concern.[9] It was physics and not biology that was the natural science of greatest political concern.

Once more the imperial bureaus played a role in internal Soviet politics. The main purpose of *The New Genetics in the Soviet Union* (Hudson and Richens 1946) was to inform Western scientists about the Soviet genetics controversy. But it also became part of the background material for the Central Committee's decision in 1948, and was known at least in summary to Lysenko himself. It was used by Iu. Zhdanov in preparing his fateful lecture against Lysenko in April 1948 (Krementsov 1997, 153; Babkov 1998, 89–90).

The purpose of P. S. Hudson and R. H. Richens was to "describe the principal tenets of Lysenko's school and to determine as far as possible their validity."[10] Patiently and dispassionately, the dialectical materialist background of the new genetics was explained, including the "historical and psychological" context. The book does present the main reasons why Lysenko's claims against classical genetics have to be rejected: he argues from authority and philosophical principles rather than from scientific evidence, and the experiments he refers to are not reliable. Thus "Lysenko's rejection of the data accumulated by Mendelian genetics during the past thirty years is obscurantist and reduces the value of his speculations," concluded Hudson and Richens (1946, 76). However, the critical impact of the book depended very much upon the reader's own ability to understand the weakness of Lysenko's scientific claims. The account was likely to produce a positive impression of Lysenko's work in readers who had little or no scientific training. There is no doubt that Hudson and Richens were well aware of the difference in scientific status between Lysenko's plant physiology and his genetics.

But the introduction of their book nevertheless sets Vavilov, Michurin, and Lysenko on par as scientists who all three have rendered great services to world genetics.[11] In a review in *Nature* Eric Ashby praised the book for its sober and balanced account in "the best tradition of scholarship," but he did also note that Hudson and Richens were "almost orientally apologetic every time they bring out a verdict against" Lysenko (Ashby 1946, 285–87).

But the Society for Freedom in Science kept warning about the threat of Lysenkoism. In an editorial *Nature* commented on an article from the society and a letter from Bernal, who criticized the journal's editorials on "conditions of survival" for being anti-Soviet. Both contributions were published in the same issue of the journal. The editorial defended a middle course. On the one hand, some planning of science was good, provided "it is the man of science who must be allowed to do the planning in consultation with others." On the other hand, Bernal was rebuked for suggesting that anyone who criticized Soviet general or science policy was a victim of the late Goebbel's propaganda. "We think it is Bernal who is allowing politics to intrude upon his scientific views, and this is the type of attitude which we feel must be checked."[12]

There was optimism in the West about the development of Soviet genetics up to 1948. Ashby wrote in 1946 that "it is safe to assume that Lysenko's school has passed its zenith" (Ashby 1946, 287). The optimism appeared to be confirmed by Zhebrak (1945) and Dubinin (1947) in *Science*. Western scientists did not take note of the ominous signs accompanying the advent of the cold war. They were not prepared for the radical reassertion of the party line in genetic science during the summer of 1948.

EXPLAINING THE RISE OF LYSENKOISM

It is important to note that Lysenko's genetics was abandoned long before the fall of the Soviet regime. By the early 1960s the identity and chemical structure of DNA as the primary hereditary material had been established. The green revolution was gaining momentum:

agriculture was being transformed worldwide by modern genetics. And Nikita Khrushchev visited Iowa to learn about the wonders of hybrid corn. In this international context it was simply impossible to hold on to Lysenko's genetics anymore. Loren Graham has argued that this final refutation of Lysenkoism demonstrates the futility of a radical constructivist interpretation of science. He wants a more fruitful approach in science studies, a sophisticated constructivism that accommodates social influence as well as the "intrusion of nature" at the core of scientific knowledge formation (Graham 1998, 17–31). This argument suggests that the *rise* of Lysenkoism must be an interesting case for developing such a sophisticated constructivism. Why was the Soviet system not able to perceive and act on the dysfunctional character of Lysenkoism long before 1960?

Zhores Medvedev has pointed to a lack of free public debate on scientific issues and a centralized system of government in science as major factors. Besides the self-censorship of people under constant threats of terror and political reprisals, there was an elaborate system of formal censorship. This largely eliminated public debate about scientific issues of some political sensitivity. From about 1934 to 1964, for thirty years the central press (*Pravda* and *Izvestia*) did not accept serious articles criticizing Lysenkoism, though many were submitted. This lack of criticism reinforced the standardization of research owing to ideological pressure and centralized bureaucratic steering. A more pluralistic science system open to criticism and decentralized in decision making would be much more resistant to the spread of false doctrines such as Lysenkoism (Medvedev 1969, 249–52).

However, it was not simply because the system of censorship and centralized government of science ceased to exist that Lysenko's genetics was finally rejected. The system continued to exist. But administrative suppression could not stop the discussion but only force it underground. In the end, the arguments made an impression on the country's leadership in spite of the repressive system of science politics. Long before the system disappeared, Medvedev wrote that the change in central government attitude and "the subsequent gradual elimination of Lysenkoism in our country are connected with the mighty scientific patriotism of public opinion, which little by little was formed among Soviet scientists of all disciplines, among journalists and writers, and among public figures and directors of the national economy" (Medvedev 1969, 253). Scientific criticism succeeded in spite of the repressive system.

Medvedev's explanation expresses a deep and sincere enlightenment belief in science as a progressive cultural force. For him it was the true Marxist intellectual heritage that finally managed to get rid of a false doctrine and set a derailed Soviet biology right. In this respect Medvedev's perspective is different from that of the recent institutional historians of Stalinist science (Krementsov 1997; Kojevnikov 2004), who focus on rituals and "games" in organizations and governing bodies through which status, power, and resources are distributed. Questions about the truth or falsity, rationality or irrationality, of scientific claims do not appear as essential driving elements in their accounts. From a traditional enlightenment attitude, this appears as a somewhat cynical view of science.[13]

David Joravky's account in *The Lysenko Affair* (1970) is moralistic. It analyzes how the evils of the Stalinist system affected an area of science. Scientists are good or bad, or humanely weak, in the face of evil political influence and coercion. Joravsky gives little room for serious scientific debate and genuine disagreements over classical genetics are given little attention. The expectations of fundamental change found in Vavilov and Sapegin, and the attempts of Zhebrak to find a mechanism for Lamarckian inheritance, are passed over in silence. In general, Joravsky neglects the uncertainties and open issues that faced science as well as the politics of science at the time. He makes the scientific superiority of classical genetics too obvious. As a result, there is much unwarranted Whig wisdom of hindsight in his account of the Soviet genetics controversy. Joravsky's moral rejection of the Stalinist system is unobjectionable, but his cavalier treatment of the scientific and science policy debates, ignoring the scientific seriousness of the opposition to classical genetics, is unsatisfactory.

Loren Graham similarly makes superficial and misleading judgments about good and bad science, for instance, with respect to Lysenko's early work in plant physiology. He does not analyze the content in its contemporary scientific context. Lysenko's 1928 article

"The Influence of the Thermal Factor on the Duration of the Phases of Development of Plants" (Graham 1987, 105–107) is dismissed without looking into, for instance, his contacts with the cotton specialist Gavril Zaitzev, a good friend of Vavilov. With such distance to the scientific issues, it is easy to miss the strength of valid scientific support for parts of Lysenko's work. This applies, for instance, to the reasons for Vavilov's enthusiastic support of vernalization research. Lysenko's methods of vernalization offered a solution to Vavilov's problems in making effective practical use of the Wold Collection. Valery Soyfer (1994, 53–57) mentions this point, but does not develop its implications. Vavilov's openness to Lysenko's positive contributions, in spite of his serious flaws as a scientist, is apparently difficult for a hardened anti-Lysenkoist fighter like Soyfer to appreciate.

I have claimed that genuine scientific issues and genuine scientific results had a more fundamental formative role in the development of Lysenkoism than the political explanations of Joravsky and Graham admit. To better grasp the *interdependence* of science and politics I have analyzed the contents of the science policy debates. On this basis I have challenged both Graham's view that Lysenkoism had little to do with serious a Marxist theory of science (1972, 195) and Joravsky's that "Lysenko's genetics originated entirely apart from intellectual processes of the community of biologists" (Joravsky 1970, 207). Joravsky does not explain well why the answer to the agricultural problems was Lysenko and not, for instance, Vavilov. The reason for this costly mistake was not just the willful bossism of the top leadership. A necessary condition was the system of science administration, politics, and advice that had been built with the active and partly enthusiastic cooperation of the scientific community.

Two important philosophical themes shaping the intellectual framework of Soviet biology were Lamarckism and a pragmatic conception of scientific knowledge often expressed by Marxists as the "unity of theory and practice."

LAMARCKISM

Lamarckism has an affinity to biological worldviews that stress the dependence of living organisms on their environment. In holistic views of biology the inextricable interdependence of organism and environment is often emphasized for individual development as well as the evolution of species. The inheritance of acquired characters was a formula that was easy to understand. It also appeared morally responsible to those who wanted a more egalitarian and just society. It seemed natural that plants that were thriving due to good husbandry would transmit some of their vitality to the offspring, and it appeared just as fair if those who spoiled their bodies in debauchery were punished by transmitting a reduced quality of inheritance to their offspring. Belief in the heredity of syphilis, tuberculosis, and other diseases was widespread. And alcohol was widely seen as a poison that contributed to hereditary degeneration. In the late nineteenth and early twentieth centuries there was strong synergy between socialist, puritan, temperance/abolitionist, and eugenics movements.

It took extensive and conscientious genetic research to show that Lamarckian beliefs were in general untenable. The new experimental biology demonstrated again and again that Mendelian heredity and natural selection of arbitrary variation provided the best explanations of hereditary and evolutionary phenomena. But where neo-Darwinism faced strong cultural and political or religious opposition, it often failed to convince. In fact, it is still seriously challenged by Christian creationism in countries that are leading in biological science. And in the recent holistically oriented philosophy of biology, the clear distinction between genotype and phenotype that was so important to early classical genetics is becoming blurred. Heredity is located not only in structures containing the standard hereditary material of nucleic acids, but in any cellular structure with some persistence in transmission from one generation to the next, and even in stable features of the environment (Sterelny and Griffiths 1999).

The history of Lysenkoism has often been seen from a somewhat dogmatic and narrow perspective of classical genetics without considering the uncertainty about Mendelism and chromosome theory among the broader community of Soviet biological scientists. The Lamarckian sympathies of Vladimir Liubimenko, nestor of plant physiology, or Vladimir Komarov, president of the Academy of Sciences, was more than just personal scientific prejudice or lacking knowledge of contemporary research. It represented a well-founded tradition of scientific doubt and critical attitude toward a young and overconfident special discipline.

One reason for the lasting affinity between Marxism and Lamarckian theories of biological heredity was probably that Marxism, as a secular religion, needed a "meaningful" world picture, and Lamarckian theories satisfied the human desire for "meaning" better than their neo-Darwinian alternatives. For instance, it placed a responsibility on parents for the heredity of their offspring that neo-Darwinism relieved them of.

Lysenko could draw on a strong Lamarckian tradition in Russian biology. Daniel Todes (1989) has argued that strictly selectionist Darwinian views of evolution had less support in nineteenth-century Russian biology than in the West. In contrast to the Western tradition, there was a wish in Russia to emphasize cooperation, mutual aid, and common striving rather than the harsh competition of Darwinian natural selection. Most serious Lamarckian biologists did not cooperate with Lysenko or directly support his ideas. In this narrow sense, they "had nothing to do with the birth of Lysenkoism" (Joravsky 1970, 207). It is also true that they, like most other scientists, became more and more negative toward Lysenko as his unscientific methods of experimenting and arguing were revealed. But they supported ideas of the Lamarckian type that were in conflict with classical genetics and in harmony with Lysenkoism. In this way they contributed to a general openness to Lysenko's ideas. Since the discussion was heavily slanted toward abstract theoretical and philosophical questions, this widespread sympathy for Lamarckism was an important source of legitimation for Lysenko.[14]

We have seen that especially at one crucial moment, this vague sympathy for Lamarkian views had a strong influence on fateful decisions, namely, when Stalin put his authority and political clout behind Lysenko in the summer of 1948.

IMPACT OF MARXIST PHILOSOPHY OF SCIENCE

I have argued that *belief* in the practical usefulness of Lysenko's results was the primary reason for the Soviet government's support of his research. This does not imply that it was farming rather than theoretical ideology that was at issue. Joravsky has emphasized how the Stalinist invocation of practicality, the so-called practice criterion of the truth of theories, was in itself a powerful ideology. It was "an extreme version of that characteristic triumph of 'pragmatism' in the twentieth century" (Joravsky 1983, 575). I have shown how discussions in the Lenin Academy of Agricultural Science (VASKhNIL) during the crucial years in the mid-1930s were dominated by the principle of unity of theory and practice, ending up with the parody of "socialist competition" in scientific research. The slogan *Ne otstavat' ot zhizny!* (Don't Hang Back from Life!) pointed to the primacy of (practical) politics over (theoretical) science. It was indeed an extreme and aggressive version of the view that "science should be on tap not on top," popular also in Western politics (Price 1954, 129).

The ideological framework of Soviet science policy was developed especially by Aleksandr Bogdanov and his pupil, Nikolai Bukharin. Bogdanov's radical empiricist monism, inspired by Ernst Mach, took a very critical view of all kinds of "metaphysics," including the theories of natural science. Theories were instruments for action rather than descriptions of reality. This was a philosophy of science that suited the Stalinist practice criterion of truth very well. It was in a deep sense a social constructivist theory of science. Though Lenin in his famous *Materialism and Empiriocriticism* sharply criticized Bogdanov's "empiriocriticism" for being idealist, his attitude was ambiguous, and Bogdanov's influence remained strong.[15] Bogdanov led a vigorous *proletkult* (proletarian culture) movement in the years immediately following the October revolution. The movement was soon suppressed by Lenin, but its ideas revived when

Stalin let loose the cultural revolution of 1928–1932. Bukharin developed his visions of a new Soviet kind of science based in the experiences of the working masses rather than in the observations and experiments of scientists. The content of his doctrines for science policy was largely a product of pre-Stalinist thinking. But Stalinism added important emphasis and constraints. The demand for *partiinost'* (party-mindedness) drastically narrowed the room for criticism. Objections were interpreted as political disloyalty, and the countermeasures became draconian (Barber 1981). The soft social and humanistic sciences were hard hit, and the natural sciences were also soon under heavy pressure (Joravsky 1961).

After a relaxation of the ideological pressure in 1932, new strictures came with the purges and terror in 1936 and 1937. The liquidation of the old guard of Bolsheviks and the rise of a new Soviet intelligentsia provided a cultural environment that suited Lysenkoism. In the arts a sophisticated modernism was suppressed in favor of simplistic socialist realism. Hardheaded and simple solutions were at a premium. There was a general tendency to return to more primitive attitudes toward knowledge and reason (Fitzpatrick 1978, 1979). While the 1936 VASKhNIL congress on breeding and genetics was still primarily a scientists' conference, the 1939 conference organized by the party's theoretical journal, *Under the Banner of Marxism*, was dominated by philosophers. In 1936 the principles of Marxist theory of science were applied by a leading scientist, Georgii Meister (1937), who gave the official evaluation of the results of the congress. In 1939 this job was done by a philosopher (Mitin 1939). The same philosopher also played a central role in preparing the 1948 congress.

The aspirations of Marxist philosophy to hegemony over science continued past the fall of Lysenko in 1965. I. T. Frolov, an influential philosopher of liberal political inclinations, defended the role of the philosophers in the 1939 conference in his book *Genetics and Dialectics* (1968). He claimed that the aim of Mark Mitin and other philosophers at the 1939 conference was a viable compromise between the two directions in genetics, taking a position similar to that of Vavilov in 1936 (Frolov 1968, 82). The philosophers, according to Frolov, were calling for "friendly cooperation and mutual principled criticism of insufficiencies and mistakes, for getting rid of the one-sidedness of the theoretical positions to be defended." It was only later that this attitude changed, and that growing intolerance led to the 1948 congress (Frolov 1968, 97). However, Frolov's comparison of Vavilov's position to that of the philosophers neglects the scientific issues. Vavilov was positive toward Lysenko's plant physiology but a consistent critic of his genetics. Frolov overlooked this essential point. He also perpetuated the scientifically superficial ideology of Michurinism by attributing to Michurin "outstanding results, which had unequaled significance for genetics and biology in general" (Frolov 1968, 45).[16]

The idea of philosophy of science as a "science of science" that can guide the politics of science had a strong hold on the Soviet academic establishment until the end. Iurii Zhdanov, the young chemist and Central Committee official who triggered the fateful showdown between Lysenko and the geneticists in 1948, expressed no criticism of the practice criterion of truth when he published and commented on his lecture forty-five years later (Zhdanov 1993).

However, more classical liberal ideas about the politics of science have also been present. Another Soviet philosopher, V. P. Filatov, in 1988 pointed to the "destruction of culture, of social-humanistic thinking" under Stalinism as the fundamental cause of Lysenkoism. Many biologists and philosophers succumbed to "proletkult spirit and vulgar sociological rhetoric," and in particular the idea of a fundamental difference between "bourgeois" and "socialist" science. In Filatov's view the geochemist and academician V. I. Vernadsky gave the appropriate answer to these tendencies when during the hard times he stressed the international validity of science across national and political boundaries. Thus Filatov chose as his hero an explicit supporter of the liberal ideal of science rather than one of those geneticists, such as Zhebrak or Dubinin, who defended genetics on Marxist principles. He argued that the doctrine of two sciences is foreign to the Marxist tradition, but that this tradition was perverted toward the end of the 1920s (Filatov 1988, 13–15).

In conclusion, Filatov argues that he emergence of Lysenkoism was "conditioned by a complex interaction of sociohistoric conditions and processes." Simplistic explanations in terms of either personal ambition and conspiracy or misguided ideology are insufficient (Filatov 1988, 5).

The result of the August 1948 session of the Lenin Academy of Agricultural Science, effectively outlawing classical genetics in the Soviet Union, came as a shock to Western scientists and intellectuals. Even the geneticists with the best Soviet contacts were quite unprepared. It is only natural that the suppression of genetics in the Soviet Union has been subjected to more research than any other topic in twentieth-century history of science.

The Soviet government's wholesale rejection of a central branch of modern science is indeed hard to understand. How could a government with special allegiance to science believe that a scientific dispute could be solved by this kind of political fiat? Particularly enigmatic appears the bold decision to immediately publish a verbatim report in several international languages so anyone could read the travesty of scientific reasoning presented by Lysenko and his followers. Did the Soviet government believe so strongly in the truth of his scientific theories that it took the chance of contradicting the vast majority of the world's scientists, implying that soon Lysenko would be proven right and the triumph accordingly increased? It is thinkable that Stalin had such a belief. Frolov reports that Lysenko himself thought so to the end of his days (Frolov 1988, 14 fn). But how was it possible that the idiosyncrasy of a few persons could overwhelm a whole government?

I do not want to give an answer to the riddle of August 1948. But I believe that today the most interesting aspect of Lysenkoism is not the direct political intervention to decide a scientific issue, but how bad science grew out of good under the influence of a particular science policy. At present there is much worry about the quality of scientific and academic institutions and their output in many countries. There are crises in universities and charges of political distortion of scientific advice, for instance, on health and environmental issues. New principles of policy and organization are eagerly sought. In this situation it is worth remembering that the Soviet Union was a pioneer both in modern science politics and in mass higher education. It was an example that formed the development of science and higher education in the West through most of the twentieth century in direct as well as more subtle ideological ways. As a proverbial case of the politically induced degeneration of science, Lysenkoism is worth new attention at the beginning of the twenty-first century. The intellectual content of Lysenkoism must be set squarely into the national and international context of ideological, philosophical, and scientific discussions, as they developed at the time. In brief, it is time to study the Lysenko effect rather than the Lysenko affair.

NOTES

1. A more extensive discussion and criticism of Marxist epistemology is found in Merton (1945).
2. The Russian text: "Lysenko prav v tom, shto sovremennaia genetika otorvana ot zaprocov praktiki. Etot otryv neizbezhen, ibo gentika pazvivaetsia kak universitetskogo tipa nauka, kak nauka ob ob'ektivno sushchestestvuiushchei forme materii, a ne kaknauka o meroipriiatiiakh, napravlennikh k dostizheniiu postavlennoi sebe tseli."
3. J. D. Bernal, "Science and Society," *Spectator*, July 11, 1931. Quoted from Werskey (1978, 147).
4. "Genetic Theory and Practice in the U.S.S.R.," *Nature* 139 (January 30, 1937): 185.
5. "Genetics and Plant Breeding in the U.S.S.R.," *Nature* 140 (August 21, 1937): 296–97.
6. "A Great Soviet Biologist," column in the communist newspaper *The Daily Worker*. Reprinted in J. B. S. Haldane, *Science and Everyday Life* (London: Lawrence & Wishart, Ltd., 1939), pp. 393–94.
7. J. B. S. Haldane, "A Note on Genetics in the U.S.S.R.," *Modern Quarterly* 1 (1938): 393–94.
8. J. B. S. Haldane, *Science Advances* (London: Allen & Unwin, 1947). Subchapter on "Genetics in the Soviet Union," pp. 220–26.
9. See, for instance, the editorial on "Conditions of Survival: Freedom of Thought and the International Community," *Nature* 158 (October 5, 1946): 459–61.

10. According to Richens the book was his idea and the text was written by him. Interview with R. S. Richens in Cambridge in November 1981.

11. "It is hardly necessary to point out here the many services that Russian geneticists have already rendered to world genetics. Vavilov's phytogeographical researches on the origins of crop plants, Michurin's pioneer work on distant hybridization, and Lysenko's studies on vernalization are contributions to biology which transcend any controversial issues" (Hudson and Richens 1946, 4).

12. "Freedom in Science," editorial in *Nature* 158 (October 26, 1946): 565–67.

13. Krementsov is quite explicit about this. Before taking up the history of science, he started a career in physiology, "spending my nights experimenting in a laboratory and my days teaching in a secondary school. This experience left me profoundly disillusioned" (Krementsov 1997, xi).

14. See Gaissinovich (1980, 1988) for considerably more emphasis on the role of Lamarckism in the rise of Lysenkoism than given by Joravsky.

15. See Dominique Lecourt, "Bogdanov, Mirror of the Soviet Intelligentsia" (1977, 137–62).

16. In his 1988 book, *Filosofiia i istoriia genetiki* (Philosophy and history of genetics), Frolov does not express any change in interpretation or attitude to the role of philosophy in the 1939 genetics conference.

Abbreviations and References

ABBREVIATIONS

ARAN: Archive of the Russian Academy of Sciences.

BIa: *Biulleten' Iarovizatsiia.* Journal published by Lysenko 1932-1933.

BV: *Biulleten' VASKhNIL.* Bulletin published by VASKhNIL.

ITsKKPSS: *Itogi Tsentral'nyi Komitet Kommunisticheskoi Partiii Sovetskogo Soiuza.* Series publishing archival documents.

LG: *Literaturnaia Gazeta* (Literary Gazette).

PZM: *Pod Znamenem Marksizma* (*Under the Banner of Marxism*). Theoretical journal of the Soviet Communist Party.

SG: *Sel'skokhoziaistvennaia Gazeta* (Agricultural gazette). Daily newspaper issued by the Ministry of Agriculture. (Renamed *Sotsialisticheskoe Zemledelie* in 1930.)

SGS: Soveshchanie po genetike i selektsii, organized by the editorial board of the journal *Under the Banner of*

Marxism (PZM), October 7-14, 1939. Stenographical report exists in two volumes. I had the opportunity to consult this report at the apartment of Vassilii Babkov in Moscow.

SS: *Selektsiia i Semenovodstvo* (Selection and Seed Management).

SZ: *Sotsialisticheskoe Zemledelie* (Socialist agriculture). Called *Sel'skokhoziaistvennaia Gazeta*, SG (Agricultural Newspaper), before 1930.

SRSKh: *Sotsialisticheskaia Rekontruktsiia Sel'skogo Khoziaistva* (Socialist reconstruction of agriculture). Journal published jointly by VASKhNIL and the Ministry of Agriculture.

SVGS: *Spornye Voprosy Genetiki i Selektsii* (Disputed issues of genetics and selection). Report of IV session of VASKhNIL, December 19-27, 1936. Moscow and Leningrad: VASKhNIL, 1937.

TPBGS: *Trudy po Prikladnoi Botanike, Genetike i Selektsii* (Works in applied botany, genetics, and selction). Journal published by VIR.

TsGANTD: Tsentral'nyi gosudarstvennyi arkhiv nauchno-tekhnicheskie dela, St. Peterburg (Central State archive for scientific and technical matters, St. Petersburg).

TsGAE: Tsentral'nyi gosudarstvennyi arkhiv ekonomiki. (Central State archive for the economy), Moscow.

TVSGS: *Trudy vsesojuznogo s'ezda po genetike, selektsii, semenovodstvu i plemennomu zhivotnovodstvu, v Leningrade 10-16 janvaria 1929g* (Works of the All-Union Congress on

Genetics, Selection, Seed Production, and Animal Breeding, in Leningrad, January 10-16, 1929). 3 vols. Leningrad, 1929-30.

VASKhNIL: Vsesoiuzny Akademiia Sel'sko-Khoziaistvennykh Nauk Imeni Lenina (The All-Union Lenin Academy of Agricultural Science).

VIET: *Voprosy Istorii Estestvoznanii i Tekhniki* (Issues in the history of natural science and technology). Journal of the Institute for the History of Natural Science and Technology of the Russian (earlier Soviet) Academy of Sciences.

VIR: Vsesoiuzny Institut Rastenievodstva (All-Union Institute for Plant Industry). Located in Leningrad and directed by Vavilov until his arrest in 1940.

VL2: *Nikolai Ivanovich Vavilov, iz epistoliarnogo naslediia 1929-1940 gg. Nauchnoe Nasledstvo.* Vol. 10, *Vavilov Letters, 1929-1940.* Moscow, 1987.

REFERENCES

Adams, Mark B. 1978. "Vavilov, Nikolay Ivanovich." *Dictionary of Scientific Biography.* Vol. 15 (suppl. vol. 1), 505-13. New York: Charles Scribner's Sons.

———. 1980. "Science, Ideology, and Structure: The Kol'tsov Institute, 1900-1970." In *The Social Context of Soviet Science,* edited by L. L. Lubrano and S. G. Solomon, 173-204. Boulder, CO: Westview Press.

———, ed. 1990a. *The Wellborn Science: Eugenics in Germany, France, Brazil, and Russia.* New York: Oxford University Press.

———. 1990b. "Eugenics in Russia, 1900-1940." In Adams 1990a.

———. 1990c. "Agol, Izrail Iosifovich." *Dictionary of Scientific Biography.* Vol. 17 (suppl. vol. 2). New York: Charles Scribner's Sons.

———. 1990d. "Levit, Solomon Grigorevich." *Dictionary of Scientific Biography.* Vol. 18 (suppl. vol. 2). New York: Charles Scribner's Sons.

———. 1990e. "Levitskii, Grigorii Andreevich." *Dictionary of Scientific Biography*. Vol. 18 (suppl. vol. 2). New York: Charles Scribner's Sons.
———. 1990f. "Serebrovskii, Aleksandr Sergeevich." *Dictionary of Scientific Biography*. Vol. 18 (suppl. vol 2). New York: Charles Scribner's Sons.
Aleksandrov, Daniel. 1995. "The Historical Anthropology of Science in Russia." *Russian Studies in History* 34: 62–91.
Aleksandrov, V. 1987–1990. "Trudnye gody sovetskoi biologii." *Znanie-Sila* 10 (1987): 72–80; 12 (1987): 50–59; 12 (1988): 74–81; 3 (1989): 76–78; 8 (1989): 62–68; 1 (1990): 50–56; 3 (1990): 72–80.
Anon. 1932. "Puti planovogo postroeniia agrofiziologicheskikh issledovanii v SSSR." *TPBGS, Sotsialisticheskoe Rastenievodstvo*, Ser. A, no. 5–6: 3–29. (Report from the All-Union Conference on Agrophysiology, February 8–13, 1932).
Ashby, Eric. 1946. "Genetics in the USSR." *Nature* 158 (August 31): 285–87.
———. 1947. *Scientist in Russia*. Harmondsworth, England: Penguin.
Astaurov, B. L. 1941. "Pamiati Nikolaia Konatantinovicha Kol'tsova." *Priroda* 5: 109–17.
Babkov, Vasilii. 1989. "N. K. Kol'tsov: Bor'ba za avtonomiiu nauki i poiski podderzhkii vlasti." *Voprosy Istorii Estestvoznaniia i Tekhniki* 3: 3–19.
———. 1992. "N.K.Kol'tsov i evo institut v 1938–39." *Ontogenez* 4: 443–53.
———. 1998. "Kak kovalas' pobeda nad genetikoi." *Chelovek* 6: 104–11.
Bailes, K. E. 1978. *Technology and Society under Lenin and Stalin. Origins of the Soviet Technica l Intelligentsia, 1917–1941*. Princeton, NJ: Princeton University Press.
Barber, J. 1981. *Soviet Historians in Crisis, 1928–1932*. London: MacMillan.
Basova, A. P., F. Kh. Bakhteev, I. A. Kostiuchenko, and E. F. Palmova. 1935. "Problema vegatastionnogo perioda v selektsii." *Teoreticheskie osnovy* 1: 863–92.
Beliaev, E. A. 1982. *KPSS i organizatsiia nauka v SSSR*. Moscow: Izdatel'stvo politicheskoi literatury.
Bell, G. D. H. 1936. "Experiments on Vernalization." *Journal of Agricultural Science* 26: 155–71.
———. 1937. "The Effect of Low-Temperature Grain Pre-Treatment on the Development, Yield, and Grain of Some Varieties of Wheat and Barley." *Journal of Agricultural Science* 27: 377–93.
Berg, Lev S. 1926. *Nomogenesis*. London: Constable & Company, Ltd.
Bernal, John D. 1939. *The Social Function of Science*. New York: Macmillan.
———. 1949. "The Biological Controversy in the Soviet Union and Its Implications." *Modern Quarterly* 4, no. 3: 203–17.

———. 1957. *Science in History*. 2nd ed. London: Watts.
Bogdanov, Aleksandr. 1908. "Filosofiia sovremennogo estesvoispytatelia." In Bogdanov, ed. 1909, 37–142.
———, ed. 1909. *Ocherki filosofii kollektivizma, sbornik pervyi*. St. Petersburg. Contains articles by N. Verner, A. Bogdanov, B. Bazarov, A. Lunacharskii, and M. Gorkii.
———. 1918. *Sotsializm Nauki (Nauchnye zadachi proletariata)*. Moscow: Izd. zhurnala proletarskaia kul´tura.
———. 1920. *Elementy proletarskoi kyl'tury v razvitii rabochego klassa*. Moscow: Gosudarstvennoe izdatel'stvo.
Bondarenko, A. S. 1935. "Itogi oktiabrskoi sessii." *BV* 11: s. 1–3.
———. 1936. "Stakhonovskoe dvizhenie v nauke." *BV* 1: 1–4.
Bowler, Peter. 1983. *The Eclipse of Darwinism*. Baltimore: Johns Hopkins University Press.
———. 1989. *The Mendelian Revolution*. Baltimore: Johns Hopkins University Press.
Bukharin, Nikolai. 1931a. "O planirovanii nauchno-issledovatel'skoi raboty." In Bukharin 1988, 324–33.
———. 1931b. "Theory and Practice from the Standpoint of Dialectical Materialism." In Bukharin et al. 1931, 9–33.
———. 1988. *Izbrannye trudy*. Moscow: Nauka.
———. 1989. *Metodologiia i planirovanie nauki i tekhniki*. Moscow: Nauka.
Bukharin, Nikolai et al. 1931. *Science at the Cross Roads*. London: Frank Cass. Second edition 1971 with new foreword by Joseph Needham and new introduction by P. G. Werskey.
Carlson, Elof. 1981. *Genes, Radiation, and Society. The Life and Work of H. J. Muller*. Ithaca, NY: Cornell University Press.
Chailakhian, Michail. 1934. "Problema iarovizatsia rastenii za granitsei." *Priroda* 5: 72–78.
Conquest, Robert. 1986. *The Harvest of Sorrow. Soviet Collectivization and the Terror Famine*. New York: Oxford University Press.
Danilov, V. P. 1988. "Diskussiia v zapadnoi presse o golode 1932–1933 gg. i 'demograficheskoi katastrofe' 30–40-kh godov v SSSR." *Voprosy Istorii* 3: 116–21.
David, R. E. 1935. "Pshenitsa i klimat." *SRSKh* 12 (December): 123–37.
Derevitskii, N. F. 1935. "Statisticheskii metod v selektsii." *Teoreticheskie Osnovy*, 549–68.
Dobzhansky, Theodosius. 1946. "Lysenko's 'Genetics.'" *Journal of Heredity* 37: 5–9.

———. 1947. "N. I. Vavilov, a Martyr of Genetics." *Journal of Heredity* 38: 226–32.
Doel, Ronald E., Dieter Hoffmann, and Nikolai krementsov. 2003. "State Limits to International Science: A Comparative Study of German Science under the Third Reich, Soviet Science under Stalin, and US Science in the Early Cold War." Manuscript (draft dated October 25, 2003) to be published in *Osiris*.
Dolgushin, D. A., and T. D. Lysenko. 1929. "K voprosu o sushchnosti ozimi." *TVSGS*, 189–99.
Dolgushin, Ju. 1949. *U istokov novoi biologii* (At the sources of the new biology). Moscow: Toskul'tprosvetizdat.
Doroshev, I. A. 1936. "Protiv samouspokoennosti rukovoditelei s.-kh. nauchnikh uchrezhdenii." *BV* 9: 8–9.
Dubinin, Nikolai. 1929. "Genetika i neolamarckizm." *Estestvoznanie i Marxizm* 4: 75–89.
———. 1947. "Work of Soviet Biologists: Theoretical Genetics." *Science* 105 (January–June): 109–12.
———. 1975. *Vechnoe Dvizhenie*. Moscow: Politizdat.
———. 1990. *Genetika—stranitsy istorii*. Kishinev: Shtiintsa.
Dunn, L. C. 1966. *A Short History of Genetics*. New York: McGraw-Hill.
Dunn, Leslie. 1944. "Science in the USSR. Soviet Biology." *Science* 99 (January 28): 65–67.
Elina, Olga. 1995. "Nauka dlia sel'skogo khoziastva v Rossiskoi Imperii: formy patronazha." *Voprozy instorii estestvoznaniia i tekhniki* 1: 40–65.
———. 1997a. "Dionisy Rudzinsky, the Plant Breeding Station at the Moscow Agricultural Academy, and Its Contacts with Svalöf, 1900-1917." *Journal of the Swedish Seed Association* 107: 225–34.
———. 1997b. "Selskokhoziaistvennye stantsii v nachale 1920-kh: Sovetskii variant reformy" (Agricultural research stations in the beginning of the 1920s: The soviet variant of reforms). In *Na Perelome: Sovetskaia Biologia v 1920-1930gg* (On the break: Soviet biology in 1920s and 1930s), 1: 27–85. St. Petersburg: Institute for the History of Science Print.
———. 2002. "Planting Seeds for the Revolution: The Rise of Russian Agricultural Science, 1860-1920." *Science in Context* 15, no. 2: 209–37.
Engledow, F. L. 1924. "The Economic Possibilities of Plant Breeding. " In *Report of the Imperial Botanical Conference, 1924*, 31–40. Cambridge: Cambridge University Press.
———. 1931. "Plant-Breeding: Its Practices and Scientific Evolution." *Scientific Journal of the Royal College of Science* 1: 74–95.

Esakov, V. 1971. *Sovetskaia nauka v gody pervoi patiletki*. Moscow: Nauka.
Esakov, V., and E. Levina. 1987. "Nikolai Ivanovich Vavilov. Pis'ma Raznyk let." *Nauka i Zhizn'* 10: 52-61.
Espinasse, P. G. 1941. "Genetics in the USSR." *Nature* 148 (December 20): 739-43.
Favorov, A. 1933. "Theoretical and Practical Significance of Lyssenko's Research on the Vernalization of Agricultural Plants." *Herbage Reviews* 1: 9-14.
Filatov, V. P. 1988. "Ob istokakh lysenkovskoi 'agrobiologii.'" *Voprosy filosofii*, 3-22.
Fitzpatrick, Sheila. 1974. "Cultural Revolution in Russia, 1928-1932." *Journal of Contemporary History* 9: 33-52.
———, ed. 1978. *Cultural Revolution in Russia, 1928-31*. Bloomington: Indiana University Press.
———. 1979. "Stalin and the Making of a New Soviet Elite." *Slavic Review* 38.
Frolov, I. T. 1968. *Genetika i dialektika*. Moscow: Nauka.
———. 1988. *Filosofiia i istoriia genetiki, poiski i diskussii*. Moscow: Nauka.
Gaissinovich, Abba E. 1973. "Problems of Variation and Heredity in Russian Biology in the Late Nineteenth Century." *Journal of the History of Biology* 6: 97-123.
———. 1977. "Problemy istorii genetiki v sovetskikh publikatsiiakh poslednykh let." *Genetika* 13: 345-71.
———. 1980. "The Origins of Soviet Genetics and the Struggle with Lamarckism, 1922-1929." *Journal of the History of Biology* 13: 1-51.
———. 1985. "Contradictory Appraisal by K. A. Timiriazev of Mendelian Principles and Its Subsequent Perception." *History and Philosophy of the Life Sciences* 7: 257-86.
———. 1988. *Zarozhdenie i razvitie gentiki*. Moscow: Nauka.
Gaissinovich, A. E., and K. O. Rossianov. 1989. "'Ia gluboko ubezhdion chto ia prav. . . .'" *Priroda* 5: 86-95, 6: 95-103.
Garner, W. W., and H. A. Allard. 1920. "Effect of the Relative Length of Day and Night and Other Factors of the Environment or Growth and Reproduction in Plants." *Journal of Agricultural Research* 18: 553-606.
Gassner, Gustav. 1918. "Beiträge zur physiologischen Charakteristik sommer-und winterannueller Gewächse, insbesondere der Getreidepflanzen." *Zeitschrift für Botanik* 10: 417-80.
Gatovskii, S. 1935. "Na odesskoi sessii Akademii S.-Kh. Nauk im. Lenina." *SRSKh* 8, no. 2 (August): 205-10.
Gibbons, Michael, et al. 1994. *The New Production of Knowledge. The Dynamics of Science and Research in Contemporary Societies*. London: Sage.

Glass, B., W. Gajewski, A. Putrament, S. M. Gershenson, and R. L. Berg. 1990. "The Grim Heritage of Lysenkoism: Four Personal Accounts." *Quarterly Review of Biology* 65: 413–79.

Glushchenko, I. E. 1937. " Uchiony iz narod—T. D. Lysenko." *Biologiia v shkole* 5: 30–46.

Goldschmidt, R. B. 1949. "Research and Politics." *Science* 109 (March 4): 219–27.

Graham, Loren. 1967. *The Soviet Academy of Sciences and the Communist Party, 1927–1932*. Princeton, NJ: Princeton University Press.

———. 1972. *Science and Philosophy in the Soviet Union*. New York: Alfred A. Knopf.

———. 1987. *Science, Philosophy, and Human Behavior in the Soviet Union*. New York: Columbia University Press. (An expanded, updated, and revised version of Graham 1972.)

———. 1993. *Science in Russia and the Soviet Union. A Short History*. New York: Cambridge University Press.

———. 1998. *What Have We learned about Science and Technology from the Russian Experience*. Stanford, CA: Stanford University Press.

Gregory, R. A. 1921. "The Message of Science." *Science* 54 (November 11): 447–56.

———. 1923. *Otkrytiia, tseli i znachenie nauki* (Petrograd). Translation from English edition *Discovery, or the Spirit and Service of Science*. London 1917, under the editorship of N. I. Vavilov.

Grille, D. 1966. *Lenins Rivale, Bogdanov und seine Philosophie*. Köln: Verlag Wissenschaft und Politik.

Guliaev, G. V. et al. 1973. "Georgii Karlovich Meister." *Genetika* 9: 173–76.

Haldane, J. B. S. 1932. *The Inequality of Man and Other Essays*. London: Chatto and Windus.

———. 1940. "Lysenko and Genetics." *Science and Society, A Marxian Quarterly* 4: 433–37.

———. 1949. "In Defence of Genetics." *Modern Quarterly* 4, no. 3: 194–202.

Harland, S. C. 1948. "The Lysenko Controversy. Four Scientists Give Their Points of View." *Listener* (London) (December 9): 873.

Harman, Oren Solomon. 2003. "C. D. Darlington and the British and American Reaction to Lysenko and the Soviet Conception of Science." *Journal of the History of Biology* 36: 309–52.

Harris, D. R. 1990. "Vavilov's Concept of Centers of Origin of Cultivated Plants: Its Genesis and Its Influence on the Study of Agricultural Origins." *Biological Journal of the Linnean Society* 39: 7–16.

Hartshorne, E. Y. 1937. *The German Universities and National Socialism.* Cambridge, MA: Harvard University Press.
Harwood, Jon. 1984. "The Reception of Morgan's Chromosome Theory in Germany." *Medizinhistorisches Journal* 19: 3–32.
———. 1985. "Genetics and Evolutionary Synthesis in Interwar Germany." *Annals of Science* 42: 279–301.
———. 1993. *Styles of Scientific Thought. The German Genetics Community, 1900–1933.* Chicago: University of Chicago Press.
———. 1997. "The Reception of Genetic Theory among Academic Plant-Breeders in Germany, 1900–1930." *Journal of the Swedish Seed Association* 107: 187–95.
Harwood, W. S. 1906. *The New Earth.* New York: Macmillan. Russian translation, V. S. Garvud, 1909, *Obnovlionnaia zemlia*, Moscow.
Hudson, P. S. 1936. "Vernalization in Agricultural Practice." *Journal of the Ministry of Agriculture* , 536, 543.
Hudson, P. S., and R. H. Richens. 1946. *The New Genetics in the Soviet Union.* Cambridge, England: School of Agriculture.
Hunter, H. 1939. "Developments in Plant Breeding." In *Agriculture in the Twentieth Century. Essays on Research, Practice, and Organization to Be Presented to Sir David Hall*, 223–60. Oxford: Clarendon Press.
Huxley, Julian. 1949. *Soviet Genetics and World Science.* London: Chatto and Windus.
Iakovlev, Ja. A. 1937. "O darvinisme i nekotorykh antidarvinistakh" (On Darwinism and some anti-Darwinist). *SRSKh* 4 (April): 17–26.
Jablonka, Eva, and M. J. Lamb. 1995. *Epigenetic Inheritance and Evolution. The Lamarckian Dimension.* Oxford: Oxford University Press.
Johannsen, Wilhelm. 1903. *Über Erblichkeit in Populationen und in reinen Linien.* Jena: Gustav Fischer.
———. 1909. *Elemente der exakten Erblichkeitslehre.* Jena: Gustav Fischer.
Jones, G. 1979. "British Scientists, Lysenko and the Cold War." *Economy and Society* 8: 26–58.
Joravsky, David. 1961. *Soviet Marxism and Natural Science.* London: Routledge & Kegan Paul.
———. 1963. "The First Stage of Michurinism." In *Essays in Russian and Soviet History*, edited by J. S. Curtiss, 120–32. Leiden: E. J.Brill.
———. 1965. "The Vavilov Brothers." *Slavic Review* 24: 381–94.
———. l970. *The Lysenko Affair.* Cambridge MA: Harvard University Press.
———. 1978. "The Construction of the Stalinist Psyche." In Fitzpatrick, ed. 1978.

———. 1983. "The Stalinist Mentality and the Higher Learning." *Slavic Review* 42: 575–600.
———. 1989. *Russian Psychology*. Oxford: Basil Blackwell.
———. 1997. "Struggles to Beat the System." *Nature* 385: 783–84.
Josephson, Paul R. 1981. "Science and Ideology in the Soviet Union: The Transformation of Science into a Direct Productive Force." *Soviet Union* 8: 159–85.
———. 1988. "Science Policy in the Soviet Union, 1917–1927." *Minerva* 26: 342–69.
Keller, Boris. 1920. "Ekologiia rastenii v eia otnosheniiakh k genetike i selektsii." In *Trudy III vserossiiskogo s'ezga po selektsii i semenovodstve v g. Saratove iiun' 4–13, 1920,* 68–74. Saratov, 1920.
———. 1933. "Otvet akad. Kellera tov. Petrovu." *Semenovodstvo* 5 (September–October): 55–56.
———. 1936. "Genetika i Evoliutsiia." *SRSKh* 12 (December): 23–32.
Kitcher, Philip. 2001. *Science, Truth, and Democracy*. New York: Oxford University Press.
Klebs, Gustav. 1910. "Alterations in the Development and Forms of Plants as a Result of Environment." *Proceedings of the Royal Society of London*, series B, 82: 547–58.
Kholodny, Nikolai. 1939. "The Internal Factors of Flowering." *Herbage Reviews* 7: 223–47.
Kojevnikov, Alexei. 1996. "President of Stalin's Academy. The Mask and Responsibility of Sergei Vavilov." *Isis* 87: 18–50.
———. 1998. "Rituals of Stalinist Culture at Work: Science and the Games of Intraparty Democracy circa 1948." *Russian Review* 57: 25–52.
———. 1999. "Dialogues about Knowledge and Power in Totalitarian Political Culture." *Historical Studies in the Physical and Biological Sciences* 30: 227–47.
———. 2004. *Stalin's Great Science*. London: Imperial College Press.
Kolakowski, Lezek. 1978. *Main Currents of Marxism*. 3 Vols. Oxford: Clarendon Press.
Kolbanovskii, V. 1939. "Spornye voprosy genetiki i selektsii." *PZM* 11: 86–126.
Kolchinskii, E. I. 1991. "Nesostoiavshiicia 'soiuz' filosofiia i biologii (20-30-e gg.)." In *Repressirovannaia Nauka,* 34–70. Leningrad: Nauka.
Kol'tsov, Nikolai.1937a. "Uspekhi sovetskoi nauki v oblasti biologii za dva desiatiletiia." *Biologicheskii Zhurnal* 6: 929–46.
———. 1937b, contribution to the discussion, *SVGS,* 237–43.
Komarov, Vladimir. 1937. "Izdavat' Timiriazeva i Darvina." *SRSKh* 4 (April): 27–29.

———. 1944. *Uchenie o vide u rastenii. Moskva*. Leningrad: Izdatel'stvo Akademiia Nauk SSSR.

———. 1945. *Izbrannye Sochineniia*. Moscow: Izdatel'stvo Akademii Nauk SSSR.

Konstantinov, P. N. 1937. Contribution to discussion. *SVGS*, 187–205.

Konstantinov, P. N., P. I. Lisitsyn, and D. Kostov. 1936. "Neskol'ko slov o rabotakh odesskogo instituta selektsii i genetiki." *SRSKh* 11 (November): 121–30.

Korol, A. G. 1965. *Soviet Research and Development: Its Organization, Personnel, and Funds*. Cambridge, MA: Harvard University Press.

Krementsov, Nikolai. 1996. "A 'Second Front' in Soviet Genetics: The International Dimension of the Lysenko Controversy, 1944–1947." *Journal of the History of Biology* 29: 229–50.

———. 1997. *Stalinist Science*. Princeton, NJ: Princeton University Press.

Krenke, N. P. 1937. contribution to the discussion. *SVGS*, 301–11.

Lakatos, Imre. 1971. "History of Science and Its Rational Reconstructions." In *Philosophical Papers*, vol.1, edited by Imre Lakatos, 102–38. New York: Cambridge University Press.

Lang, Anton. 1956. "Michurin, Vavilov, and Lysenko." *Science* 124 (August 10): 277.

———. 1957. "Genetics, Corn, and Potato in the USSR." *Plant Science Bulletin* (a publication of the Botanical Society of America, Inc.) 3, no. 3 (July): 1–2.

Lassan, Tania. 1997. "The Bureau of Applied Botany." *Journal of the Swedish Seed Association* 107: 221–24.

Lecourt, Dominique. 1977. *Proletarian Science? The Case of Lysenko*. London: New Left Books.

Levina, E. S. 1991. "Tragediia N. I.Vavilova." In *Repressirovannaia Nauka*, 223–39.

Levitskii, G. A. 1937. Contribution to the discussion. *SVGS*, 153–55.

Lewontin, Richard, and Richard Levins. 1976. "The Problem of Lysenkoism." In *The Radicalization of Science. Ideology of/in the Natural Sciences*, edited by H. Rose and S. Rose, 32–64. London: MacMillan.

Liovshin, L. V. 1987. "Bratia Vavilovy." *Priroda* 10 (October): 108–14.

Lisitsyn, Piotr. 1935. "Semenovodstvo i sortosmena." *SRSKh* 12 (December): 148–54.

———. 1937. Contribution to discussion. *SVGS*, 160–63.

Liubimenko, Vladimir. 1923. *Kurs Obshchei Botaniki*. Berlin: R.S.F.S.R. Gosudarstvennoe izdatel'stvo.

———. 1924. *Biologiia Rastenii*. Moscow: Gosudarstvennoe izdatel'stvo.

———. 1933. "K teorii iskusstvennogo regulirovaniia dliny vegetatsionnogo perioda u vyshikh rastenii" (On the theory of artificial regulation of the vegetation period in different plants). *Sovetskaia Botanika* 6: 3–30.

———. 1937. "Plant Physiology in the USSR." *Plant Physiology* 12: 895–97.

Liubishchev, A. A. 1991. *V zashchitu nauki* (To the defence of science). Leningrad: Nauka.

Lysenko, Trofim D. 1923. "Tekhnika i metodika selektsii tomatov na Belotserkovskoi selekstantsii" (Technique and method in the selection of tomatoes at the Belotserkov selection station). *Biulleten Sortovodnosemennogo Upravleniia*, part 4 (Kiev): 73–76.

———. 1928. *Vliianie termicheskogo faktora na prodolzhitel'nost' faz razvitiia rastenii, opyt co zlakami i khlopchatnikom, Trudy Azerbaidzhanskoi Tsentral'noi Opytno-Selektsionnoi Stantsii, im. tov. Ordzhonikidze v Gandzhe.* Vypusk 3, Baku.

———. 1929. "Vliianie termicheskogo faktora na fasy razvitiia u rastenii i programma rabot po etomu voprosu so sviokloi." *Trudy Vsesoiuznogo Tsentral'nogo Nauchno-issledovatel'skogo Instituta Sakharnoi Promyshlennosti (TsIS)* Moscow, 2: 34–36. (Report from a scientific meeting, December 12–19, 1928, on research connected with the sugar industry).

———. 1932a. "Resul'taty opytov 1930 goda s iarovizovannymi posevami v kolkhozakh i sovkhozakh USSR." *BIa* 1 (January): 57–61.

———. 1932b. "Predvaritel'noe soobshchenie o iarovizirovannykh posevakh pshenits v sovkhozakh i kolkhozakh v 1932g." *BIa* 2–3 (August–September): 3–15.

———. 1932c. "Osnovnye resultaty rabot po iarovizatsii sel'sokhozaisvennykh rastenii." *BIa* 4 (October–December): 3–59.

———. 1932d. "K voprosu o regulirovanii dliny vegetatsionnogo perioda s.-kh. Rastenii." *BIa* 1 (January): 5–13.

———. 1934. "Fiziologija razvitija rastenij v selektsionnom dele." *Semenovodstvo* 2 (February): 20–31.

———. 1935a. "O perestrojke semenovodstva." *Iarovizatsiia* 1 (July–August).

———. 1935b. "K stat'e neskol'ko kriticheskikh zamechanii akad. G. K. Meistera." *SS* 2, no.10 (October): 21–30.

———, ed. 1935c. *Teoreticheskie osnovy iarovizatsii*. Moscow-Leningrad: Sel'khozgiz.

———. 1936a. "Naverstat' poteriannyi god." *Iarovizatsiia* 2–3 (March–June): s. 9–18.

---. 1936b. "Proverka praktikoi." *Selektsiia i semenovodstvo* 12 (December): 5-9.

---. 1936c. *Teoreticheskie osnovy iarovizatsii*, 2nd ed. Moscow: Sel'khozgiz.

---. 1936d. "Otvet na statiu 'Neskol'ko slov o rabotakh odesskogo instituta selektsii i genetiki' akad. Konstantinova P.N., akad. Lisitsyna P.I. i Doncho Kostova." *SRSKh* 11: 131-38.

---. 1936e. "O vnutrisortovom skreshchevanii rastenii-samoopylitelei." *SRSKh* 10 (August): 70-86.

---. 1937. "O dvukh napravleniiakh v genetike." *Iarovizatsiia* 1 (January-February): s. 31-75.

---. 1939. "Genetics in the Soviet Union. Three Speeches from the 1939 Conference on Genetics and Selection." *Science and Society, a Marxian Journal* 4: 183-233. Lysenko's speech on pp.196-218.

---. 1946. *Heredity and Its Variability*, translated by Th. Dobzhansky. New York: Kings Crown Press.

---. 1954. *Agrobiology, Essays on Problems of Genetics, Plant Breeding, and Seed Growing*. Moscow: Foreign Languages Publishing House.

Lysenko, Trofim D., and A. S. Okonenko. 1923. "Privivka sakharnoi sviokly." *Biulleten Sortovodno-semennogo Upravleniia Sakharotresta*, part 4 (Kiev): 77-80.

Lysenko, Trofim D., and I. I. Prezent 1935. *Selektsiia i teoriia stadiinogo razvitiia rasteniia*. Moscow: Sel'khozgiz.

Maksimov, Nikolai A. 1929a. "Fiziologicheskie sposoby regulirovaniia dliny vegetatsionnogo perioda." *Trudy Vsesoiuznovo s'eszda po genetike, selektsii i semenovodstvo* (Leningrad) 3: 1-19. A footnote tells that a full version will be printed in *TPBGS*, presumably Maksimov 1929b below.

---. 1929b. "Fiziologicheskie faktory, opredeliaiushchie dlinu vegetatsionnogo perioda." *TPBGS* 20: 169-209.

---. 1929c. "Eksperimentelle Änderungen der Länge der Vegetationsperiode bei der Pflanzen." *Biologisches Zentralblatt* 49: 513-43.

---. 1932. "Puti rekonstruktsii fiziologii rastenii." *TPBGS, Ser. A. Sotsialisticheskoe rastenievodstvo* 2: 32-41.

---. 1934. "The Theoretical Significance of Vernalization." *Imperial Bureaus of Plant Genetics Bulletin* 16 (December): 14.

---. 1947. "Fiziologii rastenii." In *Ocherki po istorii russki botaniki*, edited by Breslavets et al., 211-27. Moscow: Moskovskoe obshchestvo ispytatelei prirody, 1947.

Maksimov, Nikolai A., and M. A. Krotkina. 1929-30. "Issledovaniia nad

posledeistviem ponizhionnoi temperatury na dlinu vegetatsionnogo perioda." *TPBGS* 23, no. 2: 427–78.
Maksimov, Nikolai A., and A. I. Pojarkova. 1925. "Über die physiologische Natur der Unterschiede zwischen Sommer- und Wintergetreide." *Jahrbücher für wissenschaftliche Botanik* 64: 702–30.
Manevich, E. A. 1990. *Such Were the Times*. Northhampton, MA: Pittenbruch Press.
———. 1991. "V zashchitu N. I. Vavilova." *VIET* 2: 138–43.
Manoilenko, K. V. 1969 *Ocherki iz istorii izucheniia fitogormonov v otechestvennoi nauke*. Moscow: Nauka.
———. 1996. *V. N. Liubimenko: evolutsionnye, ekologo-fiziologicheskie, istorikonauchnye aspekty deiatel'nosti*. St. Petersburg: Nauka.
Mather, Kenneth. 1942. "Genetics and the Russian Controversy." *Nature* 149 (April 18): 427–30.
McGucken, W. 1978. "On Freedom and Planning in Science: The Society for Freedom in Science, 1940–46." *Minerva* 16: 42–72.
McKinney, H. H. 1940. "Vernalization and the Growth-Phase Concept." *Botanical Review* 6: 25–47.
McKinney, H. H., and W. J. Sando. 1933a. "Russian Methods for Accelerating Sexual Reproduction in Wheat." *Journal of Heredity* 24: 165–66.
———. 1933b. "Earliness and Seasonal Growth Habit in Wheat." *Journal of Heredity* 24: 169–79.
McKinney, H. H., et al. 1934. "Field Experiments with Vernalized Wheat." United States Department of Agriculture, Washington, DC, circular no. 325.
Medvedev, Roy A. 1980. *Nikolai Bukharin: The Last Years*. New York: W. W. Norton.
Medvedev, Zhores. 1967 "U istokov geneticheskoi diskussii." *Novyi Mir* 4: 226–34.
———. 1969. *The Rise and Fall of T. D. Lysenko*. New York: Columbia University Press.
———. 1978. *Soviet Science*. New York: W. W. Norton.
———. 1998. "Lit' vodu na mel'nitsu zhebrakov . . ." *LG* (July 15): 3.
Meister, Georgii. 1927. "Problema mezhvidovoi gibridizatsii v osveshchenii sovremennogo eksperimental'nogo metoda." *Zhurnal opytnoi agronomii iugo-vostoka* 4: 3–86.
———. 1934. *Kriticheski ocherk osnovnykh poniatii genetiki*. Moscow: Gosudarstvennoe izdatel'stvo po kolkhoznoi I sovkhoznoi literatury.
———. 1935a "Neskol'ko kriticekikh zamecanii." *SS* 2, no. 10 (October): 13–17.

———. 1935b. "Sovjetskaia selektsiia, Jeio dostizeniia i perspektivy." *SRSKh* 12 (December): 138–47.
———, ed. 1936. *Posobie po selektsii*. Moscow.
———. 1937. "Itogi diskussii po voprosam genetiki i selektsii." *BV* 1: s. 4–19.
Merton, Robert K. 1938. "Science and the Social Order." Reprinted in Merton 1973, 254–66.
———. 1942. "Science and Technology in a Democratic Order." Reprinted as "The Normative Structure of Science" in Merton 1973, 267–78.
———. 1945. "Paradigm for the Sociology of Knowledge." Reprinted in Merton 1973, 7–40.
———. 1973. *The Sociology of Science. Theoretical and Empirical Investigations*. Chicago: University of Chicago Press.
Mitin, Mark B. 1939. "Za peredovuiu sovetskuiu nauku." *PZM* 10: 149–76.
Morgan, T. H., A. H. Sturtevant, H. J. Muller, and C. B. Bridges. 1915. *The Mechanism of Mendelian Heredity*. New York: Henry Holt.
Morton, A. 1952. *Soviet Genetics*. London: Lawrence and Wishart.
Moskovskaia sel'skokhoziaistvennaia akademiia K.A.Timiriiazeva. 1965. Moscow: Kolos.
Mosolov, V. P. 1935. "Printsipy postroeniia agrotekhniki psenitsy v usloviiakh severnoi necernozemnoi polosy." *SRSKh* 12, no. 6 (December): 176–87.
Muralov, A. I. 1935a. "Nauka i sotsialisticheskoe zemledelie." *BV* 4: 12–18.
———. 1935b. "Itogi pervoi sessii." *BV* 5: 1–4.
———. 1935c. "Speech to the October 1935 Meeting of VASKhNIL." *SS* 4, no. 12 (December): 13–22. Also printed as an editorial in *Pravda*, October 27, 1935.
———. 1935d. "Itogi oktiabrskoi sessii Akademii S.-Kh. Nauk im. Lenina." *SRSKh* 12, no. 6 (December): 98–109.
———. 1936a. "Sotsialisticheskoe sorevnovanie na vyshuiu stupen.'" *BV* 12 (December): 1–5.
———. 1936b. "Sel'skokhoziaistvennaia nauka na putiakh perestroiki." *BV* 9: 3–7.
———. 1936c. "Zadachi dekabr'skoi sessii." *BV* 12: 1–3.
———. 1936d. "Vstupitel'noe slovo prezidenta akad. A. I.Muralova." *Biulleten' IV Sessii* 1 (December): 2–3.
———. 1937. "Osnovnye itogi IV sessii." *BV* 1: 1–3.
Murinov, A. D. 1913. "Koloshenie ozimykh rzhi i pshenitsy pri iarovom poseve." *Zhurnal opytnoi agronomii* 14: 238–54.

———. 1914. "K biologii ozimykh khlebov. Koloshenie ozimykh rzhi i pshenitsy pri iarovom poseve." *Iz resul't. vegetats. op. i lab. M. S.-Kh. Inst.* 9: 167–252.

Murneek, A. E. and R. O. Whyte, eds. 1948. *Vernalization and Photoperiodism.* Waltham, MA: Chronica Botanica Company.

Needham, Joseph. 1938. "Genetics in the USSR." *Modern Quarterly* 1: 370–74. Published under the pseudonym "Helix and Helianthus."

Nerling, O. 1933. "Die Jarowisation des Getreides nach T. D. Lysenko." *Der Züchter* 5: 61–67.

Nuttonson, M. Y. 1948. "Some Preliminary Observations of Phenological Data as a Tool in the Study of Photoperiodic and Thermal Requirements of Various Plant Material." In Murneek and Whyte 1948, 129–43.

Orlovskii, N. V. 1980. *Aleksei Grigor'evitch Doiarenko.* Moscow: Nauka.

Ostwald, Wilhelm. 1909. *Grosse Männer.* Leipzig: Akademische verlagsgesellschaft m.b.h.

Palladino, Paolo. 1994. "Wizards and Devotees: On the Mendelian Theory of Inheritance and the Professionalization of the Agricultural Science in Great Britain and the United States, 1880–1930." *History of Science* 32: 409–44.

Pangalo, K. I. 1931. "Selektsiia, eio razvitie i znachenie v narodnom khoziaistve." *Priroda* 1: 40–56 (columns).

Paul, Diane. 1979. "Marxism, Darwinism, & The Theory of Two Sciences." *Marxist Perspectives* 2: 116–43.

———. 1983. "A War on Two Fronts: J. B. S. Haldane and the Response to Lysenkoism in Britain." *Journal of the History of Biology* 16: 1–37.

Pavlukhin, Ju. S. 1994. "Nikolai Aleksandrovich Maksimov." In *Soratniki Nikolai Ivanovich Vavilova*, 347–63. St. Petersburg: VIR.

Pechnikova, S. S. 1937. "Sushchnost' i zadachi sovetskoi fenologii." *Sovetskaia Botanika* 3: 105–106.

Pesikov, Ju. 1987. "O chiom napomnil staryi snimok." *LG* 25 (November): 12.

Petrov, S. 1933. "O knige akad. Kellera 'Genetika.'" *Semenovodstvo* 3 (May–July): 43–45.

Pisarev, V. 1925. "Der gegenwärtigen Zustand der Pflanzenzüchtung in Russland." *Zeitschrift für Pflanzezüchtung* 10: 221–53.

———. 1929. "Fortschritte auf dem Gebiete der angewandte Botanik und Pflanzenzüchtung in der Union der Sozialistischen Sowjetrepublikken während der letzten zehn Jahre." *Zeitschrift für Pflanzezüchtung* 14: 175–232.

Polanyi, Michael. 1943. "The Autonomy of Science." *Memoirs and Proceedings of the Manchester Literary and Philosophical Society* 85: 19–38.
Popovskii, A. 1948. *Iskusstvo tvoreniia*. Moscow: Profizdat.
Popovskii, Mark. 1966. "Tysiacha dnei Akademika Vavilova." *Prostor* (Alma Ata) 7 and 8: 4–27, 98–118.
———. 1983. *Delo akademika Vavilova*. Ann Arbor: Hermitage.
———. 1984. *The Vavilov Affair*. Hamden, CT: Archon Press.
Prezent, I. I. 1931. "Problema nauchnykh kadrov v osveshchenii burzhuaznogo biologiia. K voprosu o partiinosti nauki." *PZM* 6: 160–77.
———. 1932. *Klassovaia bor'ba na estestvenno-nauchnom fronte*. Moscow-Leningrad: Gosudarstvennoe uchebno-pedagogicheskoe izdatel'stvo.
———. 1936. "Biologiia razvitiia rastenii i genetika." *Iarovizatsiia* 1 (January–February): 3–46.
Price, Don. 1954. *Government and Science. Their Dynamic Relation in American Democracy*. New York: New York University Press.
Printz, H. 1965. "Litt om fenologi" (Something about phenology). *Blyttia* (Journal of the Norwegian Botanical Society) 23: 1–20.
Razumov, V. I. and R. N. Griuntukh. 1936. "Voprosy iarovizatsii v inostrannoi literature." *Iarovizatsiia* 2–3, nos. 5–6: 187–206.
Repressirovannaia Nauka. 1991. Edited by M. G. Iaroshevskogo. Leningrad: Nauka.
Revenkova, A. I. 1962. *Nikolai Ivanovich Vavilov, 1887–1943*. Moscow: Izdatel'stvo sel'skokhoziastvennoi literatury.
Reznik, Semion. 1968. *Nikolai Vavilov*. Moscow: Molodaia gvardiia.
———. 1981. *Zaveshchanie Gavrila Zaitseva* (The testament of Gavril Zaitsev). Moscow: Detskaia literatura.
———. 1983. *Doroga na eshafot*. New York: Tretiia Volna.
Richens, R. H. 1969. "The Commonwealth Bureau of Plant Breeding and Genetics." *University of Cambridge School of Agriculture Memoirs* (Cambridge) 41, Review Series no. 24.
Rodionov, A. D., and F. I. Filatov. 1936. "Itogi iarovizirovannykh posevov zernovykh v kolkozakh SSSR v 1935 godu." *Iarovizatsiia* 1: 82–108.
Roll-Hansen, Nils. 1978. "The Genotype Theory of Wilhelm Johannsen and Its Relation to Plant Breeding and Evolution." *Centaurus* 22: 201–35.
———. 1980. "Eugenics before World War II. The Case of Norway." *History and Philosophy of the Life Sciences* 2: 269–98.
———. 1984. "E. S. Russel and J. H. Woodger—The Failure of Two Twentieth-Century Opponents of Mechanist Biology." *Journal of the History of Biology* 17: 399–428.

———. 1985. *Ønsketenkning som Vitenskap. Lysenkos innmarsj i Sovjetisk Biologi 1927–37.* Oslo: Universitetsforlaget.

———. 1987. "Zhizn' i trudy Vavilova v zapadnoi literature." *Voprosy Istorii Estestovoznaniia i Tekhniki* 4: 52–56.

———. 1989. "The Crucial Experiment of Wilhelm Johannsen." *Biology and Philosophy* 3: 303–29.

———. 1990. "Croisement de lignées pures: de Johannsen á Nilsson-Ehle." In *Histoire de la genetique, Pratiques, Techniques et Theories,* edited by J.-L. Fischer and W. Schneider, 99–125. Paris: ARPEM.

———. 1997. "The Role of Genetic Theory in the Success of the Svalöf Breeding Program." *Journal of the Swedish Seed Association* (Sveriges Utsädesförenings Tidskrift) 107: 196–207.

———. 2000. "Theory and Practice: The Impact of Mendelism on Agriculture." *C. R. Acad. Sci. Paris, Sciences de la vie/Life Sciences* 323: 1107–16.

Rossianov, Kirill O. 1991. "Stalin kak redaktor Lysenko. K predistoriia avgustovskoi (1948 g.) sessii VASKhNIL." Preprint, Moscow, 1991.

———. 1993. "Editing Nature. Joseph Stalin and the 'New' Soviet Biology." *Isis* 84: 728–45.

Ruse, Michael. 1975. "Woodger on Genetics: A Critical Evaluation." *Acta Biotheorethica* 24: 1–13.

Sakharov, Andrei. 1990. *Memoirs.* New York: Alfred A. Knopf.

Sapegin, A. A. 1932. "Die züchterische Bedeutung der Verkürzung der Vegetationsperiode nach T.D. Lysenko." *Der Züchter* 4: 147–51.

———. 1935. "Znachenie iarovizatsii dlia fitoselktsii." In *Teoreticheskie Osnovy,* 807–18.

Sapp, Jan. 1987. *Beyond the Gene. Cytoplasmic Inheritance and the Struggle for Authority in Genetics.* New York: Oxford University Press.

Sax, Karl. 1944. "Soviet Biology." *Science* 99 (April 14): 298–99.

Sbornik diskusionnikh statei po voprosam gentiki i selektsii. 1936. Moscow: VASKhNIL.

Schliemann, Elisabeth. 1939. "Gedanken zur Genzentrentheorie Vavilovs." *Naturwissenschaften* 27: 377–83, 394–401.

Schimper, A. F. W. 1898. *Pflanzen-Geographie auf physiologischer Grundlage.* Jena: G. Fischer.

Serebrovskii, Aleksandr. 1926. "Teoriia nasledstvennosti Morgana i Mendelia i marxisty." *PZM* 1: 98–117.

Shatskii, A. L. 1930. "K voprosu o summe temperatur, kak sel'skokhoziaistvenno-klimaticheskom indekse." *Trudy po sel'sko-khoziaistvennoi meteorologii* 21, no. 6: 259–67.

Shaw, George B. 1949. "The Lysenko Muddle." *Labour Monthly* (January): 18-20.
Sinskaia, E. N. 1937. Contribution to the discussion. *SVGS*, 252-55.
——. 1991. *Vospominaniia o N.I.Vavilove*. Kiev: Naukova Dumka.
The Situation in Biological Science. 1949. Proceedings of the Lenin Academy of Agricultural Sciences of the USSR. Session: July 31-August 7, 1948. Verbatim report. Moscow: Foreign Languages Publishing House.
Skripchinskii, V. V. 1969. "Ocherk istorii fisiologii razvitiia rastenii." In *Problemy fisiologii rastenii*, edited by P. A. Genkel and E. M. Senchenkova, 45-87. Moscow: Nauka.
Solomon, S. G. 1977. *The Soviet Agrarian Debate. Controversy in Social science, 1923-1929*. Boulder, CO: Westview Press.
Soyfer, Valerii. 1989. *Vlast' i nauka. Istoriia razgroma genetiki v SSSR*. Tenafly, NJ: Hermitage.
——. 1994. *Lysenko and the Tragedy of Soviet Science*. New Brunswick, NJ: Rutgers University Press.
Spornye voprosy genetiki i selektsii. Raboty IV sessii VASKhNILA 19-27 dek. 1936 g. Moscow, VASKhNIL, 1937.
Stalin, V. I. 1929. "K voprosam agrarnoi politiki v SSSR." *Sochineniia* (Collected Works) 12: 141-72. Speech at conference for Marxist agronomist, December 27, 1929, reprinted from *Pravda*, December 29, 1929.
Stebutt, A. 1913. "Der Stand der Pflanzenzüchtung in Russland." *Zeitschrift für Pflanzenzüchtung* 1: 7-58.
Sterelny, Kim, and Paul E. Griffith. 1999. *Sex and Death. An Introduction to Philosophy of Biology*. Chicago: University of Chicago Press.
Takhtadzhian, A. 1987. "Kontinenty Vavilova." *LG* (November 25): 12.
Talanov, V. V. 1924. *Selektsiia i Semenovodstvo v SSSR*. Moscow: Novaia derevnia.
Teoreticheskie osnovy selekt´sii rastenii (Theoretical foundations of the selection breeding) of plants). 1935. Vol. 1, *Obshchaia selektsiia rastenii*, edited by N. I. Vavilov. Moscow, Leningrad: Sel'khozgiz.
Thomas, K. 1971. *Religion and the Decline of Magic. Studies in Popular Beliefs in Sixteenth- and Seventeenth-Century England*. London: Weidenfeld and Nicolson.
Tiumiakov, N. 1967. "Zabytie stranitsy otechestvennoi selektsii." *Sel'skokhoziaistvennoe proizvodstvo povolzhia* 8 (August): 24-28.
Todes, Daniel. P. 1989. *Darwin without Malthus. The Struggle for Existence in Russian Evolutionary Thought*. New York: Oxford University Press.
Tolmachev, M. I. 1929. "K voprosu o fiziologicheskoi prirode stebeleobrazovaniia u ozimnei i sakharnoi svioklovitsy." *TVSGS*, 539-53.

Tsitsin, N. V. 1937. Contribution to the discussion. *SVGS*, 293-97.
Tucker, R. C. 1992. *Stalin in Power. The Revolution from Above, 1928-1941.* New York: W. W. Norton.
Tulaikov, N. M. 1935. "Voprosy agrotekhniki psenitsy na cernozemnykh i kashtanovykh pochvakh." *SRSKh* 12 (December): 188-96.
Vasil'ev, I. M. 1934. "O faktorakh iarovizatsii ozimykh rastenii." *Doklady Akademiia Vauk SSSR* 3: 533-39.
Vavilov, Iurii. 1998. "Avgust 1948. Predistoriia." *Chelovek* 4: 104-11.
Vavilov, Nikolai. 1922. "The Law of Homologous Series in Variation." *Journal of Genetics* 12: 47-89.
———. 1926. *Tsentry proizkhozhdeniia kulturnykh rastenii.* Leningrad: Vsesoiuznyi institute prikladnoi botanki. Contains extensive English translation, "Studies on the Origin of Cultivated Plants."
———. 1930a. "(VASKhNIL) i eio osnovnye zadachi." *SZ* 22 (January): 2.
———. 1930b. "Liuter Berbank (1829-1926)." Introduction to *Zhatva Zhizni*, by L. Berbank and V. Kholl, 7-14. Moscow and Leningrad: Sel'khozgiz. Translation of Vilbur Hall, *Harvest of Life.*
———. 1931a. "Agronomicheskaia nauka v usloviiakh sotsialisticheskogo khoziaistva." *SRSKh* 5-6: 72-73.
———. 1931b. "The Problem of the Origin of the World's Agriculture in the Light of the Latest Investigations." In Bukharin et al. 1931, 95-106.
———. 1931c. "The Linnean Species as a System." *Fifth International Botanical Congress, Cambridge, August 16-23, 1930, Report of Proceedings*, 213-16. Cambridge: Cambridge University Press, 1931.
———. 1932a. "The Process of Evolution in Cultivated Plants." *Proceedings of the Sixth International Congress of Genetics, Ithaca, New York, 1932*, vol. 1, 331-42. New York: Brooklyn Botanical Gardens.
———. 1932b. "Genetika na sluzhbe sotsialisticheskogo zemledeliia." *TPBGS*, Ser. A, no. 4.
———. 1934a. "Sovetskoe nauchnoe rastenievodstvo za period sotsialistcheskoi rekonstruktsii 1930-1933." *TPBGS, Sotsialisticheskoe Rastenievodstvo* 10: 5-23.
———. 1934b. "Osnovnye zadatchi sovjetskoi selektsii rastenii i puti ikh osushchestvleniia." *TPBGS, Sotsialisticeskoe Rastenievodstvo* 12: 5-22.
———. 1935a. "Shest' let raboty Akademii s.-kh. nauk im. V.I. Lenina." *BV* 6: 15-20.
———. 1935b. "Pshenitsa v SSSR i za granitsei." *SRSKh* 12 (December): 110-22.
———. 1935c. "Organizator pobed severnogo zemledeliia." *BV* 1: 3-4.

———. 1936. "Puti sovetskoi rastenievodcheskoi nauki." *SRSKh* 12 (December): 33–46.

———. 1937. "Puti sovetskoi selektsii." *SRSKh* 2 (February): 27–49.

———. 1939. "Genetics in the Soviet Union." *Science and Society, A Marxian Quarterly* 4: 183–233. Vavilov's speech on pp. 184–96.

Vucinich, A. 1980. "Soviet Physicists and Philosophers in the 1930s: Dynamics of a Conflict." *Isis* 71: 236–50.

———. 1984. *Empire of Knowledge. The Academy of Sciences of the USSR (1917–1970)*. Berkeley: University of California Press.

Warming, E. 1895. *Plantesamfund. Grundtræk af den økologiske Plantegeografi*. Copenhagen: P. G. Philipson. (Soon translated into German.)

Weiner, D. 1984. "Community Ecology in Stalin's Russia, 'Socialist' and 'Bourgeois' Science." *Isis* 75: 684–96.

———. 1998. *Models of Nature: Ecology, Conservation, and Cultural Revolution in Soviet Russia*. Bloomington: Indiana University Press.

Werskey, P. G. 1971. "Introduction" in Bukharin et al. 1971, xi–xxix.

———. 1978. *The Visible College*. London: Allen Lane.

Wetter, Gustav. 1958. *Dialectical Materialism: A Historical and Systematic Survey of Philosophy in the Soviet Union*. London: Routledge & Kegan Paul.

Wheatcroft, Stephen. 1984. "A Reevaluation of Soviet Agricultural Production in the 1920s and 1930s." In *The Soviet Rural Economy*, edited by Robert C. Stuart. London: Rowman & Allanheld.

Whyte, R. O. 1939. "Phasic Development of Plants." *Biological Reviews of the Cambridge Philosophical Society* 14: 51–87.

———. 1948. "History of Research in Vernalization." In Murneek and Whyte 1948, 1–38.

Whyte, R. O., and P. S. Hudson. 1933. "Vernalization or Lyssenko's Method for the Pretreatment of Seed." *Imperial Bureaus of Plant Genetics Bulletin* (Aberystwyth) 9 (March).

Young, Robert. 1978. "Getting Started on Lysenkoism." *Radical Science Journal* 6–7: 80–105.

Zaitsev, Gavril. 1927. "Vliianie temperatury na razvitie khlopchatnika" (The effect of temperature on the development of the cotton plant). *Trudy turkestaskoi selektsionnoi stantsii*, vypusk 7. (Transactions of the Turkstan Plant Breeding Station, Tashkent, issue 7). Reprinted in Zaitsev 1963.

———. 1963. *Izbrannye Sochineniia*. Moscow: Izdatel; stvo sel'skokhoziaistvennoi iteratury.

———. 1980. *Izbrannye Trudy*. Tashkent: "FAN" Uzbekskoi SSR.

Zavadovskii, B. M. 1937. "Za perestroiku geneticheskoi nauki." *SVGS*, 163-83.

Zavadovskii, M. M. 1936. "Protiv zagibov v napadkakh na genetiky." *SRSKh* 8 (August): 84-95.

———. 1937. contribution to the discussion, *SVGS*, 398-405.

Zhdanov, Iurii. 1993. "Vo mgle protivorechii" (In the darkness of contradictions). *Voprosy Filosofii* 7: 65-92.

Zhebrak, Anton.1936a. "Nekotorye sovremennye voprosy genetiki" (Some present problems in genetics). *SRSKh* 8 (August): 97-122.

———. 1936b. "Kategorii genetiki v svete dialekticheskogo materializma" (Categories of genetics in the light of dialectical materialism). *SRSKh* 12 (December): s. 78-88.

———. 1945. "Soviet Biology." *Science* 102 (October 5): 357-58.

Zirkle, Conway, ed. 1949. *Death of a Science in Russia*. Philadelphia: University of Pennsylvania Press.

———. 1959. *Evolution, Marxian Biology, and the Social Scene*. Philadelphia: University of Pennsylvania Press.

Index

Academy of Sciences, 226, 231, 233, 238, 262
Adams, Mark, 204, 264
Agassiz, Louis, 60
Agol, Israil, 195, 215n, 231
agricultural economics, 90, 91–92
Aleksandrov, V., 279
Alikhanian, S. I., 267
Allard, H. A., 30, 133
All-Russian Institute of Applied Botany and New Crops. *See* VIR (All-Union Institute of Plant Industry), origins of
All-Union Institute for Genetics and Plant Breeding, Odessa, 104, 115, 118, 120, 160
All-Union Institute of Plant Industry. *See* VIR (All-Union Institute of Plant Industry)
Angewandte Botanik (journal), 148
applied science. *See* science, "pure" (basic, theoretical) vs. "applied" (practical)
Ashby, Eric, 69, 130, 151n, 290
atheism, 86–87

atomic bomb, 268, 289
autonomy of science, 85, 156, 207–208, 243, 282, 284–86

Baker, John, 286
Bakh, A. N., 263
basic science. *See* science, "pure" (basic, theoretical) vs. "applied" (practical)
Bateson, William, 40
Bauer, Julius, 232
Bauman, K. Ia., 195, 214, 239
Baur, Erwin, 28, 64, 225
Belaia Tserkov experimental station, 39, 53–55, 71
Berg, L. S., 211, 258
Berlin Blockade, 273
Bernal, J. D., 18n, 41, 78, 243, 286, 290
Biffen, R. H., 40, 144
Biulleten' Iarovizatsiia (journal), 122, 124, 126, 143
Bogdanov, Aleksandr, 79–80, 256, 296
Bondarenko, A. S., 98, 104, 125, 172

326 | INDEX

Bonnevie, Kristine, 232
Botanicheskii Zhurnal (journal), 275
branched wheat, 271–72
British Genetics Society, 240
Bukharin, Nikolai, 77, 78–84, 218, 286, 296–97
Burbank, Luther, 26, 44–45, 49
Bureau of Applied Botany. *See* VIR (All-Union Institute of Plant Industry), origin of

Central Committee, 105, 213, 221, 253, 265, 273, 274
    report on VASKhNIL, 267
Chadwick, Sir David, 145
Chailakhian, M. K., 38, 147
chromosome theory, 43, 47, 176, 201
classical genetics, 12, 21–23, 48, 142, 155, 295
    criticism of, 158–59, 164, 182–84, 187, 194, 206–207
    See also Mendel, Gregor
collectivization, 76, 90–92
Commissariate of Agriculture. *See* Ministry of Agriculture
Commonwealth Bureau of Plant Breeding and Genetics. *See* Imperial Bureau of Plant Genetics (Plant Breeding)
Communist Academy, 85, 87, 109n, 134
Conference on Agrophysiology, 1932, 136–39
Conference on Drought Control, 1931, 44, 134
Congress of Collective Farm Shock Workers, 1935, 97
Congress on Genetics, Plant Breeding, Seed Production, and the Raising of Pedigreed Livestock, Leningrad 1929, 64
Correns, Carl, 24
Council of Ministers, 96–97, 233
"court of honor," 268
cultural revolution, 81, 92, 149

Darlington, Cyril, 288
Darwin, Charles, 23, 24, 198–99, 209, 273
Darwinism, 80, 210, 218–19
    "creative Darwinism," 86–89
Deborin, A. M., 84–86, 88
Decree on Plant Breeding and Seed Production, August 1931, 133–34, 159–60, 175
Derevitskii, N. F., 61, 71n
Der Züchter (journal), 143
Detskoe Selo, Experimental Station for Genetics and Plant Breeding, 37, 38, 94
developmental physiology. *See* selection; hybridization
De Vries, Hugo, 24
dialectical materialism, 43, 83–86, 149
    in genetics, 178–84, 194, 254, 289
Dobzhansky, Theodosius, 288–89
Dubinin, Nikolai, 85–86, 110n, 258, 264–65, 269, 277n
Dunn, Leslie, 179, 235–36, 242, 265, 277n, 288–89

Ecology (journal), 147
economic efficiency, 76, 91, 118–20
ecophysiology, 29
Efroimson, V. P., 73n

INDEX | 327

Eikh'feld, I. G., 104, 132, 152n, 172, 174, 253, 264
Ekonomicheskaia Zhizn' (newspaper), 150n
Emelianenko, L., 150n
Emerson, Ralph, 110n, 234
empiriomonism (empiriocriticism), 79-80, 108n
Engels, Friedrich, 179, 182, 184, 273
Engledow, Frank, 27
Esakov, V., 71n
eugenics, 21, 87, 193, 196, 223-25, 250, 261, 286

"Fascist" race theory, 194, 232
Favorov, A., 163-64
Federley, Harry, 64, 110n, 238
Feuerbach theses (Marx), 82-83
Filatov, V. P., 298-99
Filipchenko, Iurii, 26, 87, 138, 204, 225, 258
Fischer, Eugen, 232
five-year plan, 12, 75, 92
Frolov, I. T., 297-98, 299, 301n

Gaissinovich, Abba E., 244n, 301n
Gandzha experimental station, 56, 61, 119
Garner, W. W., 30, 133
Gassner, Gustav, 30, 36, 37, 60, 65, 67, 118-19, 135, 146
gene, 21, 26-27
genotype, 25, 156-57, 260
German Society for Genetics, 225
Gibbons, Michael, 18
Glass, Bentley, 285
Goldschmidt, Richard, 64, 110n, 256
Gorbunov, Nikolai, 89, 99, 233
Gorkii, Maxim, 108n, 263

Graham, Loren, 12-13, 83-84, 291, 292-93
Gregory, Frederick, 145-46, 153n, 154n
Gregory, Richard A., 40-41

Haldane, J. B. S., 41, 78, 86, 110n, 145, 232, 242, 286-87
Harland, S. C., 173
Heide, O., 71n
Heidelberg University, 550th anniversary, 235
Hogben, Lancelot, 232
holism, 27, 50n, 164-65, 169, 183
Hudson, P. S., 144-48, 150n, 153n, 289-90
Huxley, Julian, 204, 232, 243
hybrid corn, 270, 291
hybridization, 28, 54
  choice of parent pairs, 157-58, 166-67, 169, 173, 177
  vegetative (graft), 54-55

Iakovlev, Iakov, 93-94, 98, 99, 195
  on Darwinism, 218-20
  discussion with Muller, 214
  on genetics and eugenics, 219-20
  and Lysenko, 102, 122, 134, 160
Iakovlev, P. N., 191n
Iarovizatsiia (journal), 126, 129, 151n, 174-75, 182, 218
Imperial Bureaus of Plant Genetics (Plant Breeding), 130, 143-48, 153n, 163, 261, 288-90
Imperial College, London, 145
individual selection. *See* selection
Institute for Philosophy, Academy of Sciences, 253
Institute for Selection and Genetics,

Odessa. *See* All-Union Institute for Genetics and Plant Breeding, Odessa
International Botanical Congress 1930 in Cambridge, 41
International Botanical Congress 1935 in Amsterdam, 145–46
International Commission for Agricultural Meteorology, 31
International Congress of Genetics, 1927 in Berlin, 186
International Congress of Genetics, 1932 in Ithaca, 93, 160–61, 230, 245n
International Congress of Genetics, planned for 1937 in Moscow, 93, 187, 188, 204, 225, 249
 cancellation of, 230, 233, 234
 Soviet government prohibits discussion of "race question," 237–38
International Congress of Genetics, 1948 in Stockholm, 267
International Congress of the History of Science 1931 in London, 78, 93
International Genetics Committee. *See* Permanent International Committee on Genetics
Iudin, P. F., 253, 258
Ivanov, N. R., 70n
"ivory tower," 82, 96, 285
Izvestia (newspaper), 101, 174, 196, 291

Jennings, H. S., 232
Johannsen, Wilhelm, 21, 25–26, 156–57, 165, 167, 170–71, 182, 254, 257
 Kol'man on eugenics of, 261
 rejected by Lysenko, 198–99
Joravsky, David, 12, 108n, 274, 292, 296
Journal of Heredity, 252

Kamenev, Lev, 218, 235
Karpechenko, G., 94, 110n, 185, 186–87, 188, 252
Keller, Boris A., 46, 48–49, 106
 and Lysenko, 140, 262
 and neo-Lamarckism, 48, 179
Kerkis, Iu. Ia., 255–57
Kholodny, Nikolai, 37–38, 154n
Khrushchev, Nikita, 275, 291
Kiev Agricultural Institute, 53, 56
Kirov, Sergei, 64, 131
Klebs, Gustav, 29
Koestler, Arthur, 217
Kojevnikov, Alexei, 18n
Kol', A. K., 186
Kolchisnkii, E. I., 109n
Kol'man, Arnost, 233, 253, 255–56
Kol'tsov, Nikolai, 193, 202, 204–205, 262–64
 chastized for eugenic views 223–25
 letter to president of VASKhNIL, 220, 222–23
 praising Lysenko's vernalization, 188–89
Komarov, Vladimir, 46–47, 140, 265
 on Darwinism, 219
 and International Genetics Congress, 237–38
 neo-Lamarckism, 47, 295
Konstantinov, P., 126–29, 172, 174, 189, 211–12, 220, 272

Kostov, D., 126–27, 153n
Krasnosel'ska-Maksimova, 138
Krementsov, Nikolai, 18n, 244–45n, 273, 301n
Krenke, N. P., 207–208
Kuhn, Thomas, 85
kulaks, 76

Lamarckism. *See* neo-Lamarckism
Lecourt, Dominique, 18n
Lenin, V. I., 49, 79–80, 89, 273, 296–97
Lenin Academy of Agricultural Science (VASKhNIL), 12, 17, 42, 89–92, 267–68
  aktiv of March 1937, 186, 221–30
  August 1948 congress, 273–74, 275, 299
  December 1936 congress, 193–96
  efficiency and political control, 250
  Lysenko becomes president, 250–52
  reorganization 1934–1935, 95–99
  role in collectivization, 90–92
Leningrad Botanical garden, 37
Lenin Prize, 104
Levins, Richard, 13
Levit, S. G., 204, 215n, 231, 232, 233
Levitskii, G., 94, 205–206
Lewontin, Richard, 13
liberal ideal of science, 88, 220, 223, 242–43, 282–83, 285–86, 291–92, 298
Linnaeus, Carolus, 31
Lisitsyn, P., 117, 126–28, 177, 183, 189, 269

Litvinov, M. M., 225, 238
Liubimenko, Vladimir, 37, 65, 87–88, 147
  hormone theory, 141, 145, 147
  neo-Lamarckism, 46–47, 295
Liubishev, A. A., 279n
Lobkowicz, N., 109n
London Times, 287
Lunacharskii, A. V., 108n
Lysenko, Trofim D., 11–12
  and antireductionism, 68, 164
  attacking Vavilov, 198–99, 259
  cooperation with VIR, 116, 122, 139
  death of, 276
  as director of Institute of Genetics, 265, 266
  discovering vernalization, 55, 57, 97–98
  early career of, 53–58
  at Gandzha, 56–58
  and genetic theory, 164–66, 168–71, 198–99
  holism of, 164–65, 169, 201
  on hybridization, 54–55
  at international genetics conference, 231
  and intravarietal hybridization, 228–29
  lectures at VIR, 116, 167
  as member Academy of Sciences, 263
  on Michurin, 46, 50, 169
  and Mitin, 271
  and neo-Lamarckism, 178, 200–201
  paper at 1929 Leningrad congress, 66–67
  and plant breeding, 64, 66

as president of VASKhNIL, 12, 251, 270–71
and Prezent, 140–41, 145, 167–69, 200
rejecting Mendelism, 255
and Stalin. *See* Stalin, Joseph
theory of stages, 60, 62, 66–67, 104, 142, 145, 146–47, 157, 158
treatise on temperature effects on development, 59–63
and VASKhNIL, 98, 101–102, 228, 250–52
as young enthusiast, 69–70, 174, 281
and Zaitsev, 59–62
Lysenko effect, 300
Lysenkoism, 11, 14, 17, 18, 86, 281, 291, 295–96, 298–300

Mach, Ernst, 79, 256
Makarov, N. P., 110n
Makronosov, A. T., 151n
Maksimov, Nikolai, 37
  arrest of, 94, 110n
  at Leningrad Congress 1929, 64–68
  and Lysenko 58, 65–68, 116–17, 121, 135–36, 141, 146–47
  rebuked by Vavilov, 135–36
  on vernalization, 118–19
Malenkov, Georgii, 265–66
Manevich, Eleanor, 69–70, 73n
Margolin, L. S., 186, 224–25, 229, 251
Marx, Karl, 108n, 273
  Feuerbach theses, 82–83, 283
Marxist philosophy of science, 82–84, 283, 292, 296

basis for science and science policy, 12–13, 78–84, 107, 213–14, 273–74, 296–97
effect on West, 16, 18n, 78
practice criterion, 82–83, 104, 178, 259–60, 261–62, 296
mass selection. *See* selection
McKinney, H. H., 143, 147
"mechanicists," 84–86
Mechano-Lamarckism, 88
Medvedev, Roy, 218
Medvedev, Zhores, 14, 71n, 291–92
Meister, G. K., 42–43, 65, 174, 222
  arrested, 251
  critical of Lysenko, 172, 175–77, 211–12, 250
  criticism of Vavilov, 211
  and distant hybridization, 43
  Marxist ideology, 43, 297
  summing up 1936 conference, 209–13
  and VASKhNIL, 102
Mendel, Gregor, 20, 85, 221
Merton, R. K., 109n, 282–83, 285
Michurin, Ivan, 26, 44–46, 98, 179, 254, 256
Michurinism, 46, 50
Ministry of Agriculture, 89–90, 94, 96, 99, 102, 106, 117, 134, 160, 176, 228, 254
Ministry of Health, 203
Mitin, Mark B., 253, 266, 297
  summing up 1939 conference, 253–54, 260
modernization, 76, 217
modification. *See* mutation
Mohr, O. L., 110n, 230, 232, 234, 235, 245n, 285–86
  cancellation of International

Genetics Congress, 235–37, 240–43
Molotov, V., 233, 238
Morgan, Thomas Hunt, 22, 85, 180, 196, 199, 205, 207
Morton, A., 50n
Muller, Hermann, 22, 43, 187, 193, 195–96, 239, 285
 and eugenics, 202–204, 215, 231
 and international genetics congress, 230–31, 235, 239, 241
Muralov, Aleksandr, 98, 100–102, 105–106, 178
 arrested, 251
 critical of Lysenko, 172, 251
 debating Vavilov, 227
 platform for VASKhNIL 1936 congress, 194–95
 reply to Kol'tsov's letter, 221
 summing up VASKhNIL 1936 congress, 213
Muralov, Nikolai, 218
mutation, 21, 43, 196, 210
 through "modifications," 46, 47, 48, 106, 179, 184, 200–201

Nature (journal), 40, 147, 286, 290
Navashin, M. S., 110n, 231
Needham, Joseph, 41, 78, 243, 287
neo-Darwinism, 22–23, 26, 47, 294
neo-Lamarckism, 23, 26
 in Soviet biology, 46–48, 84, 178, 295
 in Soviet politics of science, 47, 219, 238, 271, 294–95
neopositivism, 80
NEP (New Economic Policy), 75, 79, 95, 107
New York Times, 234, 235

Nikitskii Botanical Garden, 37
Nilsson-Ehle, Herman, 28, 110n, 231

orthogenesis, 47
Ostwald, Wilhelm, 56

Pasteur, Louis, 221
Pavlov, Ivan, 205
Pearson, Karl, 256
pedigree breeding. *See* selection, individual
Penrose, Lionel, 232
Permanent International Committee on Genetics, 110n, 231, 234, 237, 240, 267
Petrovka Academy for Agriculture and Forestry. *See* Timiriazev Agricultural Academy
phenology, 31–32
 and Lysenko, 59, 63
phenotype, 25, 157
philosophy and genetics, at 1939 conference, 252–53, 298
photoperiodism, 30–32, 147
Pisarev, V. E., 39, 110n, 116
planning of science. *See* science planning
plant breeding. *See* selection; hybridization
Plant Breeding Institute, Cambridge, 148
plant development. *See* plant physiology
plant physiology
 Academy of Sciences laboratory, 38
 developmental physiology of plants, 20, 28–32, 63

first Russian chair, 36
light and temperature effects,
    28–32
theory of stages (Lysenko's),
    62–64, 66–67, 142, 146–47
Pod Znamenem Marksizma. See
    Under the Banner of Marxism
    (journal)
genetics conference 1939, 253,
    297
Polanyi, Michael, 286–88
    critical of Vavilov, 288
Politburo, 272
Poltava experimental station, 115
Popovskii, Mark, 70n
Pravda (newspaper), 97, 102, 105,
    114, 115, 196, 263, 268, 272,
    291
Prezent, I. I., 70, 87–89, 152n, 263
    critical of VASKhNIL, 228–29
    and Lysenko, 138–39, 167–69
    and the suppression of genetics,
    225, 240, 250, 253, 255
Prianishnikov, Dimitrii, 33–34
    criticizing organization of
    VASKhNIL, 226–27
    and plant physiology, 36–37
Price, Don, 296
proletarian science, 79–84, 149,
    227–28
proletkult, 79, 296
pseudoscience, 13, 14
pure lines, 25–26, 197–98, 212. See
    also Johannsen, Wilhelm

Quetelet, Adolphe, 31

Razumov, V., 94, 137
Regel, Robert, 35

Reznik, S., 70n
Richens, R. H., 289, 301n
Richter, Andrei, 38, 138–39, 140,
    146
Rossianov, Kirill, 244n, 279n
Rozanova, M. A., 261
Rudzinsky, Dionisy, 35

Sabinin, D. A., 137
    criticism of Lysenko, 268
Sakharov, Andrei, 13
Sando, W. J., 143, 147
Sapegin, A. A., 110n, 157
    and Lysenko, 141–43
    on vernalization, 118, 158–59
Saratov Plant Breeding Station, 39,
    42–43, 105
Sax, Karl, 265–66
Schmalgausen, I. I., 269
Science (journal), 265
science, "pure" (basic, theoretical)
    vs. "applied" (practical), 49, 113,
    155–56, 180–82, 204, 283–85
    planning of, 78, 290
science planning, 15–16, 77, 80–
    84, 104
science politics, 50, 83, 107, 181,
    209, 229, 281–82
    Iu. Zhdanov on Lysenkoism,
    269–71
    pioneering role of Soviet Union,
    15–17, 281, 299–300
VASKhNIL *aktiv*, 226–27
VASKhNIL and collectivization,
    90–99
Zavadovskii's criticism of Soviet
    science policy, 181–82
See also "unity of theory and
    practice"

selection, 190n
 individual, 24, 54
 Lysenko and Prezent's theory and method, 167–69
 mass, 24
 as synonym for plant breeding, 155–59
Serebrovskii, Aleksandr, 84–85, 110n, 180, 202, 204, 222, 258–59
Shepilov, Dmitri, 271, 272
Shlykov, G. N., 186, 260
Shull, G. H., 45
Sinnot, E. W., 277n
Sinovjev, Grigorii, 218, 235
Sinskaia, Evgenia, 187, 207
"Socialist competition" in science, 102–107, 178, 196
social studies of agriculture, 90
Society for Freedom in Science, 287, 290
Society of Marxist Biologists, 87, 134, 140, 152n
Sotsialisticheskoie Zemledelie (newspaper), 105, 117, 218, 240
Soyfer, Valerii, 14–15, 293
Sozinov, A. A., 69, 73n
Sputnik, 16
Stakhanov movement in science, 102–104, 127
Stalin, Joseph, 195, 221
 death of, 13
 and great break, 15, 17, 75
 and Iu. Zhdanov, 270, 272, 275
 and Lamarckism, 271, 295–96
 on linguistics, 274
 and Lysenko, 97, 228, 271–72, 274, 296
 and plant breeding, 270, 271–72
 on science and collectivization, 90
 science planning and policy of, 77, 102–103, 269
Stalinist science, 13, 107, 281–82, 292, 297
Sugar Trust, 62
 experimental stations, 39–40
Svalöf plant breeding station, 19, 24–25, 42, 254

Talanov, V. V., 116
Theoretical Foundations of Plant Selection (Vavilov), 61, 155, 159
theory of stages in plant development. *See* plant physiology, theory of stages (Lysenko's); Lysenko, theory of stages
Timiriazev, Kliment, 33, 219–20, 260, 273
Timiriazev Agricultural Academy, 32–33, 265
Tincker, M. A. H., 147
Todes, Daniel, 295
Tolmatchev, M. I., 67
Tschermak, Erik von, 23
Tsitsin, N. V., 104, 152n, 208, 251–52, 262
 disagreeing with Lysenko, 267
 interspecies hybrids, 208
Tulaikov, Nikolai, 42, 43–44, 77, 89–90, 94
 arrested, 251
 critic of Vil'iams, 44
 on vernalization, 117–18, 125

Ukrainian Ministry of Agriculture, 114–15, 119–20

INDEX

Under the Banner of Marxism (journal), 87, 225
United States Department of Agriculture, 143
"unity of theory and practice," 78–79, 81–83, 95–96, 100–102, 113, 128–29, 136, 149, 156, 183–84, 193, 227, 283

Vasil'ev, I. M., 129–30
VASKhNIL. *See* Lenin Academy of Agricultural Science (VASKhNIL)
Vavilov, Nikolai, 35, 40–42, 65, 81, 110n, 184–88
  against Lysenko, 167, 254
  arrest and death of, 264, 288
  on bureucratization and scientific quality in VASKhNIL, 100, 226–27
  defending Lysenko, 171–74
  evasive toward Lysenko, 197–98
  and international genetics congress, 236, 238, 242
  and law of homologous series, 41, 185
  and Luther Burbank, 44–45
  and Lysenko, 58, 66, 70–71n, 94–95, 135–36, 160–63, 229
  at meeting in Odessa, 169–74
  and Michurin, 44–46, 186
  and philosophy of science, 81–83, 155–57
  and plant breeding as applied science, 156–57
  and plant geography 41–42
  as president of VASKhNIL, 17, 42, 77–78, 89–92, 243, 251
  twenty-fifth anniversary as scientist, 98
  and vernalization of seed grain, 149, 161, 163, 173
  and vernalization of World Collection, 162, 173
Vavilov, Sergei, 40, 265
vegetative hybridization, 54–55
Vernadskii, V. I., 87
vernalization, 31, 55, 115
  in the Arctic, 131–32
  giving up on, 129–31
  as international research topic, 133, 148–49
  origin of term, 113, 150n
  of spring grain, 119, 121, 123–24, 126, 134
  as tool for plant breeding, 134–35, 149, 158, 159–63
  of winter grain 121–22, 125
  and yields, 121, 123–26, 127–28, 146, 149
Verschuer, Otmar von, 232
Vil'iams, Vasilii, 34
VIR (All-Union Institute of Plant Industry), 58, 92–94, 105
  Arctic station at Khibiny, 132
  arrest of scientific specialists 1933, 94
  cooperation with Lysenko, 139
  and first gene bank, 41
  origins of, 35–36, 42
Vries, Hugo de. *See* De Vries, Hugo
Vucinich, Aleksander, 279n
vydvizhents, 75, 87

Weismann, August, 182
Werskey, P. G., 18
Wetter, Gustav, 108n
Whyte, R. O., 130, 144–45, 153n, 154n, 261

winter cereals, 29–30
  damage during harsh winters, 65
  vernalization of, 114–15, 117–18
Woodger, J. H., 50n, 190n
World Collection, 41–42, 97, 119, 139
  vernalization of, 158, 160, 162, 173–74

Young, Robert, 18

Zaitsev, G., 59–62
Zaitseva, M., 71n
Zavadovskii, B., 206, 224, 259
  critical of Lysenko, 273
Zavadovskii, M., 98, 102, 172, 251–52, 259, 262–63
  criticism of Prezent and Lysenko, 180–82, 206–207, 250
  and theoretical vs. applied science, 180–82, 286
Zhadanov, Andrei, 253, 268–69, 272–73
Zhdanov, Iu. A., 269–70, 272, 274–75, 298
  criticism of Lysenko, 270–71
Zhebrak, Anton R., 106, 179–80, 182–84, 252, 257–58, 265
  in Central Committee secretariate, 266–69
  and criticism of Lysenko, 183–84, 267
  on "modifications," 184
  as president of Belorussian Academy of Sciences, 267
  reply to Sax, 266
Zhukovskii, P. M., 273